Securing IoT and Big Data

Internet of Everything (IoE): Security and Privacy Paradigm

Series Editors: Vijender Kumar Solanki, Raghvendra Kumar, and Le Hoang Son

Blockchain Technology
Fundamentals, Applications, and Case Studies
Edited by E Golden Julie, J. Jesu Vedha Nayahi, and Noor Zaman Jhanjhi

Data Security in Internet of Things Based RFID and WSN Systems Applications
Edited by Rohit Sharma, Rajendra Prasad Mahapatra, and Korhan Cengiz

Securing IoT and Big Data
Next Generation Intelligence
Edited by Vijayalakshmi Saravanan, Anpalagan Alagan, T. Poongodi, and Firoz Khan

Distributed Artificial Intelligence
A Modern Approach
Edited by Satya Prakash Yadav, Dharmendra Prasad Mahato, and Nguyen Thi Dieu Linh

Security and Trust Issues in Internet of Things
Blockchain to the Rescue
Edited by Sudhir Kumar Sharma, Bharat Bhushan, and Bhuvan Unhelkar

Internet of Medical Things
Paradigm of Wearable Devices
Edited by Manuel N. Cardona, Vijender Kumar Solanki, and Cecilia García Cena

Integration of WSNs into Internet of Things
A Security Perspective
Edited by Sudhir Kumar Sharma, Bharat Bhushan, Raghvendra Kumar, Aditya Khamparia, and Narayan C. Debnath

For more information about this series, please visit: https://www.crcpress.com/Internet-of-Everything-IoE-Security-and-Privacy-Paradigm/book-series/CRCIOESPP

Securing IoT and Big Data

Next Generation Intelligence

Edited by

Vijayalakshmi Saravanan, Alagan Anpalagan,
T. Poongodi, and Firoz Khan

CRC Press
Taylor & Francis Group
Boca Raton London New York

CRC Press is an imprint of the
Taylor & Francis Group, an **informa** business

First edition published 2021
by CRC Press
6000 Broken Sound Parkway NW, Suite 300, Boca Raton, FL 33487-2742

and by CRC Press
2 Park Square, Milton Park, Abingdon, Oxon, OX14 4RN

© 2021 Taylor & Francis Group, LLC

CRC Press is an imprint of Taylor & Francis Group, LLC

Library of Congress Cataloging-in-Publication Data
Names: Saravanan, Vijayalakshmi, editor. | Anpalagan, Alagan, editor. |
Poongodi, T., editor. | Khan, Firoz, editor.
Title: Securing IoT and big data : next generation intelligence / edited by
Vijayalakshmi Saravanan, Alagan Anpalagan, T. Poongodi, Firoz Khan.
Description: First edition. | Boca Raton : CRC Press, 2021. | Includes
bibliographical references and index.
Identifiers: LCCN 2020028013 (print) | LCCN 2020028014 (ebook) | ISBN
9780367432881 (hardback) | ISBN 9781003009092 (ebook)
Subjects: LCSH: Internet of things. | Big data. | Internet of
things--Security measures. | Big data--Security measures.
Classification: LCC TK5105.8857 .S385 2021 (print) | LCC TK5105.8857
(ebook) | DDC 004.67/8--dc23
LC record available at https://lccn.loc.gov/2020028013
LC ebook record available at https://lccn.loc.gov/2020028014

ISBN: 978-0-367-43288-1 (hbk)
ISBN: 978-1-003-00909-2 (ebk)

Typeset in Times
by SPi Global, India

Contents

Preface

Technology has become unavoidable in human life, and all activities in our lives are becoming more and more interlinked with the onset of the internet, IoT, big data, and many new technologies. An abundance of data is generated by individual archives, sensors, IoT networks, social networks, and enterprises that require the exploitation of big data analytics in many aspects such as forensics, threat intelligence, and cyber security. The IoT and big data security have drawn great attention in social networks, medicine, business, transportation, stock exchanges, the energy sector, and many more.

The new paradigm shift created by big data processing in IoT delivers tailormade and secured decision support systems for the end users. The increased demand for connecting to networks and deriving the benefits of their services and demand for guaranteed privacy and security are felt more than ever and are the biggest challenge today. The number of devices present within an IoT network considerably increases the volume data about intruders that should be gathered and analysed in IoT ecosystems in terms of providing countermeasures.

This book on *Securing IoT and Big Data: Next Generation Intelligence* explores IoT threat intelligence, forensics, and cyber security challenges in big data IoT systems. The book begins with a discussion on the foundation of IoT and big data, presenting applications, and a case study in Chapter 1, followed by discussions on securing the IoT with blockchain, and the IoT and big data using intelligence in Chapters 2 and 3, respectively. Cyber intelligence for rail analytics in IORNT, and big data and IoT forensics are presented in Chapters 4 and 5. Integration of the IoT and big data in the field of entertainment for recommendation systems is discussed in Chapter 6. Privacy and security preserving data mining and aggregation in IoT applications and various data aggregation techniques are discussed in Chapter 7. Finally, Chapter 8 is devoted to healthcare IoT and a survey of a real-time cardiovascular health monitoring system using IoT and machine learning algorithms.

Acknowledgements

Vijayalakshmi Saravanan, editor-in-chief, expresses gratitude to God Almighty, her parents, and grandparents for their lifelong blessings. Special thanks go to her caring husband, Mohan Radhakrishnan, and loving daughter, Pranavi Mohan, for their relentless support, understanding, and encouragement during the writing of the book. Her sincere thanks to Dr. Anju S Pillai, Amrita Vishwa Vidyapeetham, India, for helping in editing chapters.

The editors also gratefully acknowledge the many organizations/sources for diagrams and texts referred to in the book.

Editors

Vijayalakshmi Saravanan received her PhD in Computer Science and Engineering under an Erasmus Mundus Fellowship (EURECA) as a research exchange student at Mälardalen University, Sweden, and visiting researcher at Ryerson University, Canada. She is currently working at Rochester Institute of Technology, USA. Prior to this, she was working as an Assistant Professor in Practice at the University of Texas, San Antonio (UTSA) in the Department of Computer Science. She was a Postdoctoral Associate at University at Buffalo (UB), The State University of New York, USA, and the University of Waterloo, Canada, under the prestigious "Schlumberger Faculty for the Future" Fellowship award (2015–2017). She has ten years of teaching experience in two premier universities, VIT and Amrita Vishwa Vidyapeetham, India. Dr. Saravanan has published many technical articles in scholarly international journals and conferences. She serves as a technical reviewer and programme committee member for reputed conference and journals such as GHC, SIGCSE, and Springer. Her research interests include power-aware processor design, big data, IoT, and computer architecture. She is a Senior Member of IEEE and ACM, CSI, Ex-Chair for the IEEE-WIE VIT affinity group, India (2009–2015), National Postdoctoral Association (NPA) Annual Meetings committee, Workshop/IIA Co-Chair (2017–2019), and a board member of N2WOMEN (Networking Networking Women).

Alagan Anpalagan is a Professor in the Department of Electrical and Computer Engineering at Ryerson University where he directs a research group working on radio resource management (RRM) and radio access and networking (RAN) areas within the WINCORE Lab. His current research interests include harvesting and green communications technologies, cognitive radio resource management, wireless cross layer design and optimization, cooperative communication, M2M communication, and small cell networks. Dr. Anpalagan received BSc, MSc, and PhD degrees from the University of Toronto, and also completed a course on Project Management for Scientists and Engineers at the University of Oxford CPD Centre. He is a recipient of the Dean's Teaching Award, the Faculty Scholastic, Research, and Creativity Award, and the Faculty Service Award at Ryerson University. Dr. Anpalagan is a registered Professional Engineer in the province of Ontario, Canada, a Senior Member of the IEEE and a Fellow of the Institution of Engineering and Technology.

T. Poongodi completed her PhD at Anna University, Chennai, and is currently working as an Associate Professor at Galgotias University, Greater Noida, Uttar Pradesh. She has 12 years of teaching experience in the field of computer science. Her area of interest lies in the field of IoT, big data, and networking. She has published more than ten international journals and contributed book chapters.

Firoz Khan received his PhD in Computer Science from the British University in Dubai. He is currently working in the Computer Science Department at the British University in Dubai. He has authored many research papers, contributed chapters in journals, and edited books.

Contributors

Vidushi Agarwal is currently a PhD research scholar in the Department of Computer Science and Engineering at the Indian Institute of Technology, Ropar. She was a Gold Medallist in the MTech in Computer Science and Engineering at the National Institute of Technology, Hamirpur, in 2019. Her previous research has mainly dealt with fault detection using cryptographic schemes and data integrity in wireless sensor networks. Her current areas of research interest include IoT, data security, blockchain, and wireless sensor networks.

M. Arvindhan is currently working as an Assistant Professor at Galgotias University. He obtained his bachelor's degree from Pavendhar Baharathidasan College of Engineering and Technology in 2006 and his master's degree from Prist University in 2011, and is now pursuing his PhD studies at Galgotias University. He has published various papers in national and international journals and book chapters. His area of interest is cloud computing.

Indu Chawla has been working as an Assistant Professor in the Department of Computer Science and Information Technology at Jaypee Institute of Information Technology, Noida, India. She has more than 19 years of teaching experience and has guided many graduate and undergraduate students. Her research interests are in the areas of databases, data mining, big data, and software engineering.

Vashi Dhankar is a student in the Department of Information Technology, Maharaja Agrasen Institute of Technology, Delhi. Her research interests include big data and IoT.

Wanying Dou is an undergraduate student in the Department of Computer Science, Wenzhou Kean University, China. His research interests include big data, IoT, data science, and software engineering.

R. Jothibanu is currently pursuing an MTech in Information Technology at PSG College of Technology. She completed her undergraduate studies at Anna University in 2018, having previously attended the Kurinji Institute (2011–2014). She has completed various certification courses such as C, CPP, Java, and Python. She has also gained certification from Microsoft Technologies for completing the Database Fundamentals courses.

Sujatha Krishnamoorthy is currently working as an Assistant Professor in the Department of Computer Science at Kean Wenzhou University, China, and is an active member of CSI with 19 years of teaching experience. Her area of specialization is digital image processing with image fusion. She has published over 60 papers in international refereed journals including Springer and Elsevier. She has delivered several guest lectures and seminars, and has chaired sessions at various conferences.

She serves as a reviewer and editorial board member of many reputed journals and has acted as a session chair and technical programme committee member at national and international conferences. She has received a best researcher award during her research period.

Yihang Liu is an undergraduate student in the Department of Computer Science, Wenzhou Kean University, China. His research interests include big data, IoT, data science, and software engineering.

Zixuan Liu is an undergraduate student in the Department of Computer Science, Wenzhou Kean University, China. His research interests include big data, IoT, data science, and software engineering.

D. Manojkumar is currently pursuing an MTech in Information Technology at PSG College of Technology, having previously completed his undergraduate degree at Anna University. His areas of interest include object oriented analysis and design, and software defined networking.

Mark Nuneviller is currently pursuing an MSc in Data Science at Rochester Institute of Technology, USA. His research interests include big data, data science and software engineering.

Kayal Padmanandam has over a decade of credentials in the domain of computer science with wide exposure through teaching, research, and industry. She is passionate about research and has specialized in data science and machine learning algorithms, the area in which she pursued her doctoral research. She has several publications especially related to machine learning applications. She is an ACM distinguished speaker, member of various international computing societies, and also serves as reviewer at IEEE conferences, and as an editorial team member for reputed journals. Her research interests include data science application, machine learning algorithms, artificial intelligence, data mining, IoT and big data, disruptive innovations, cloud computing, object oriented paradigms, and ethical research.

Sujata Pal received her PhD degree from the Indian Institute of Technology, Kharagpur, India. Sujata was a recipient of a Tata Consultancy Services (TCS) Research Scholarship for four years while pursuing the PhD programme. She was awarded a prestigious Schlumberger Faculty for the Future Fellowship for two consecutive years (2015 and 2016). Currently, Sujata is an Assistant Professor in the Department of Computer Science and Engineering at IIT Ropar, India. She was a postdoctoral fellow at the University of Waterloo, Canada, before joining IIT Ropar. Her research work has been published in high-quality international journals, conferences, book chapters, and a book. Her current research interests include IoT, blockchain, opportunistic mobile networks, wireless body area networks, software defined networks, and mobile ad-hoc networks.

Anju S. Pillai received her PhD in Electrical Engineering from Amrita Vishwa Vidyapeetham, India, in the year 2016. Anju was awarded the EURECA Research Fellowship, funded by the European Commission under the Erasmus Mundus external cooperation window for a period of one year at Mälardalen University, Västerås, Sweden, in 2009. She is currently an Assistant Professor (SG) in the Department of Electrical and Electronics Engineering at Amrita Vishwa Vidyapeetham, Coimbatore, India. She has more than 20 years of teaching experience. Dr. Anju has published more than 50 technical articles in scholarly international journals and conferences, and has received two best paper awards. Dr. Anju was a recipient of the Chancellor's Collaborative Research Fellowship and visited Ryerson University, Canada, in 2018 to carry out research work. Her research interests include: development of power-aware embedded systems, embedded solutions to healthcare applications, addressing challenges of multicore and multiprocessor systems, real-time scheduling, and development of fault tolerant systems.

G.S. Pradeep Ghantasala is currently working as a Professor in the Department of Computer Science and Engineering, Malla Reedy Institute of Technology and Science, Hyderabad, Telangana, India. His main research area is image processing. He has 26 published papers in reputed UGC, Scopus, and SCI journals, has presented papers at national and international conferences, and has published book chapters in CRC Press, Elsevier. He has four Indian patents. He is a guest editor/reviewer for various international journals and has given guest lectures. He received his BTech in Information Technology from JNTU, Hyderabad in the year 2006. He then pursued postgraduate studies and received his MTech degree in Computer Science and Engineering from Acharya Nagarjuna University, Guntur, and Andhra Pradesh in the year 2009. He was awarded a PhD in Computer Science and Engineering in the area of image processing in 2018 from Shri Venkateshwara University, Uttar Pradesh, India. His current research interests include the areas of image processing, image processing in medical systems, machine learning, and deep learning.

Archana Purwar has been working as an Assistant Professor at Jaypee Institute of Information Technology, Noida, India. During her teaching career of more than 14 years, she has taught subjects such as database systems, software engineering, object oriented programming, computer architecture and organization, data mining, and many more. Her area of interest lies in data mining and information retrieval.

Anu Rathee is currently working as an Assistant Professor in the Department of Information Technology, Maharaja Agrasen Institute of Technology, Delhi. Her research interests include wireless sensor networks and IoT. She has published many research papers, contributed chapters in reputed journals, and edited books.

S. Sarathambekai is currently working as an Assistant Professor in the Department of Information Technology, PSG College of Technology, Coimbatore, Tamil Nadu, India. She received her MTech in Information Technology in 2010 and completed her

PhD in 2018 in the field of distributed systems and evolutionary computation. She has published more than 30 papers in impact factored journals and conferences published by Elsevier, Springer, Wiley, Inderscience, and others. She received a Research Ratna Award (2019) for work in the area of swarm intelligence from RULA in collaboration with the World Research Council (WRC). Her areas of interest include distributed systems, swarm intelligence, web technologies, and database and algorithm design.

Emanuel Szarek is currently pursuing an MSc in Data Science at Rochester Institute of Technology, USA. His research interests include big data, data science and software engineering.

Nalli Vinaya Kumari is pursuing her PhD in Computer Science and Engineering at Sri SatyaSai University of Technology and Medical Sciences, Sehore, Bhopal. She has a BTech from Jawaharlal Nehru Technological University, Kakinada, Andhra Pradesh, India, and an MTech from Jawaharlal Nehru Technological University, Hyderabad, India. She has published various national and international papers and book chapters, and has attended conferences. Her area of interest is networking and machine learning.

K. Umamaheswari is currently a Professor and Head of Department in the Department of Information Technology, PSG College of Technology, Coimbatore, Tamil Nadu, India. She completed her bachelor`s and master`s degrees in Computer Science and Engineering during 1989 and 2000, respectively. She received her PhD in Information and Communication Engineering from Anna University in 2010. She has gained a rich experience in teaching over the last 25 years. Her area of research includes classification of various types of data such as text and images using data mining. She is also interested in the research areas of distributed scheduling, mobile communication, soft computing, and cellular automata. She has published more than 100 papers in journals. She is a recognized supervisor of Anna University. She is currently guiding four research scholars for their PhD studies and seven of her research scholars have already completed their PhD degrees. She is a life member of ISTE and ACS, and a fellow member of IE.

T. Vairam is currently working as an Assistant Professor in the Department of Information Technology, PSG College of Technology, Coimbatore, Tamil Nadu, India. She obtained her BEng in Computer Science and Engineering from Bharathiyar University in 2004 and her MTech in Information Technology from Anna University in 2008. She completed her PhD degree at Anna University in 2017. Her thesis title was "Certain Investigations on Multipath Routing in WSN for multimedia Streaming". She has published more than 20 papers in impact factored journals and conferences. She received a Research Ratna Award (2019) for work in the area of swarm intelligence from RULA in collaboration with the World Research Council (WRC). She is a lifetime member of ISTE, IEI and WRC. Her areas of interest include software defined networking and wireless sensor networks.

1 Foundation of Big Data and Internet of Things
Applications and Case Study

Vijayalakshmi Saravanan,
Rochester Institute of Technology, USA
Mark Nuneviller,
Rochester Institute of Technology, USA
Anju S. Pillai and
Amrita Vishwa Vidyapeetham, India
Alagan Anpalagan
Ryerson University, Canada

1.1 INTRODUCTION TO BIG DATA AND INTERNET OF THINGS

The healthcare sector is currently experiencing a data overload derived from the digitization revolution. There are multiple streams of data entering healthcare systems through electronic health records. Healthcare data originally was kept in the form of hard copies but following the recent trends, health systems are racing to digitize these large amounts of patient health data. The information includes medical diagnoses, prescriptions, allergy data, demographics, clinical narratives, and the clinical laboratory results [1]. With the change to digitization, we have observed improvements in healthcare systems with data being widely available, resulting in reduced examinations, fewer ambiguities caused by illegible handwriting, and fewer missing or lost patient records. Patient care has immensely improved with healthcare systems becoming more organized [2].

Electronic health records are not the only data that is being recorded; with advancements in healthcare technology, multiple instruments have transformed into digital formats. Like electronic health records, which were originally static, X-ray films, scripts, and so all went digital. Small electronics, such as heart rate monitors, EKG, and oxygen sensors, and large electronics, such as MRI machines, that are continuously recording data are all now digitized. Keeping all these streams of data organized is a serious challenge, especially with each type of data being different. The challenge of the healthcare sector is how to keep everything, including MRI images, sensory data, and patient health records, organized and interlinked. The

1

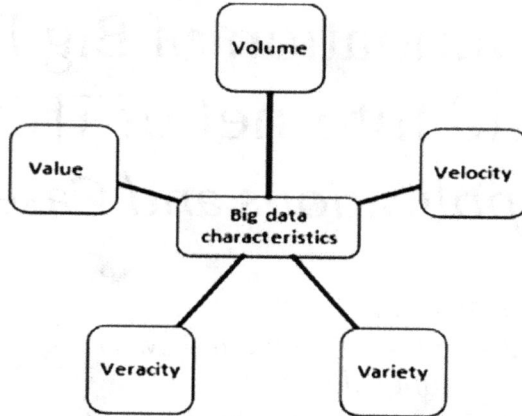

FIGURE 1.1   The Five Vs of big data.

historical relational databases that currently exist will struggle in maintaining structured and unstructured data [3].

Data analytics revolve around the five Vs of data characteristics. These are **volume**, **velocity**, **variety**, **value**, and **veracity** (shown in Figure 1.1), which will be the main challenges healthcare must overcome in order to become dominant in data organization and analytics [4]. Healthcare systems are continuously creating and accumulating data at an exponential rate. These systems need to store this large volume of data and access it at any time with a high rate of velocity. This leads to another data characteristic: the velocity that will be needed to analyse the data in real-time. Real-time monitoring of aspects such as blood pressure, operating room anaesthesia measurements, and heart monitors is crucial, as they can be the difference between life and death. Variety is a big challenge in healthcare, since the data streams come from multiple structured and unstructured sources [5]. Health systems will have to understand and manage these streams of data, such as MRI and X-ray images, sensory data, and genomic data. Healthcare applications need to be efficient in combining and converting various forms of data, including conversion from structured to unstructured data or other creative data storage solutions. Veracity is the final challenge that healthcare systems need to overcome. Is the data reliable and captured correctly? Is the data secured from outside intrusion/attack? In order to ensure these, a methodology and solution need to be constructed to tackle the challenges.

There are additional challenges in the healthcare context [6]. These include the high cost of overhauling the whole IT infrastructure that exists currently in the healthcare system. Not many healthcare systems have the money or the resources to dedicate to this kind of technological overhaul. Not only is the cost high, but also resources and a number of inputs are required to pull off a successful project. The healthcare systems will have to organize doctors, scientists, software engineers, and data scientists; all come together and bring a full technological construction, which is a huge challenge in itself.

The end goal of healthcare applications should be to successfully maintain multiple streams of data and grow to include genomic data to provide individualized healthcare treatment for patients and the highest quality of care possible. With the digitization and combining of data in order to use big data, healthcare sectors, ranging from doctors' offices to large hospital groups, can receive significant advantages. Data would be more readily available, enabling fast disease detection, so starting the right treatment within a short span of time would now be possible. These technical advancements would also lead to improvements in health status, billing fraud would be detected more quickly, and health systems would increase profits and reduce waste and redundant costs. Overall disease patterns and outbreaks would be heavily analysed to improve surveillance and speedy response. Health systems can use this data to deliver faster targeted vaccinations such as the yearly influenza strain. Data collected would turn into actionable information to meet requirements, deliver services, and both forecast and prevent crises for the benefit of mankind. Enhanced data and analytics would benefit patients, most of whom are the largest consumers of health assets and can be provided with factual and accurate information to make informed decisions, proactively handle their own health, select and track healthier practices, and identify treatments in real-time.

Genomics, not to be confused with genetics, is a new science [7]. Genomics is the study of how the entire genome (the unique genetic code that serves as a blueprint for all living things) is expressed in the physical characteristics of an organism. Genetics is an older discipline and tracks how traits are inherited from generation to generation. Genomics is young because genomes could not be sequenced until about 15 years ago. Studying the genome itself is a massive task and requires computational power that was not available until recently. One of the hopes of genomics is that it will aid in the development of drugs and treatments that are uniquely suited to a patient's genes – called precision medicine. Breakthroughs in genomics have been scarce to date, but we must bear in mind that this is a new scientific endeavour, so the going is slow but the potential rewards are huge.

1.1.1 BIG DATA MANAGEMENT SYSTEMS IN HEALTHCARE

Due to the large volume of data, health systems will have to parse and explore, and will have to rely on distributed frameworks to divide and analyse the large amount of data. Open-source projects like Hadoop and Apache Spark can be run on the cloud and provide a variety of analytics that healthcare systems can use [8, 9]. These open-source projects empower health systems to analyse large data sets that would otherwise be impossible to analyse. The data also has to be aggregated in a location where we can analyse it effectively and efficiently. Databases that are part of NoSQL technology packages, such as CouchDB and MongoDB, offer solutions to pool the raw data where it can be loaded and analysed by Hadoop, which has the ability to handle huge amounts of data with a variety of structures or no structure at all. There are multiple vendors such as Amazon Web Services (AWS), Cloudera, and MapR technologies that distribute Hadoop platforms [10, 11]. The alternative, Apache Spark, is supported by SQL and uses in-memory processing of data which makes it much faster than Hadoop but comes at a cost for large data sets. Apache Spark can be used for real-time data solutions that healthcare systems require [12].

Healthcare data also consists of signals from data such as electrocardiograms, images, and video that is stored in a patient's electronic health records. The combination of images, videos, and structured texts stored in a patient's health records can be harnessed using artificial intelligence (AI). AI programs can draw actionable insights from the wealth of structured and unstructured data to make informed decisions and diagnose diseases. Healthcare professionals can look at the abnormalities provided by these machine learning approaches.

Image analytics is important in the healthcare context, where the abundance of procedures including CT, MRI, X-ray, molecular imaging, ultrasound, PET, EEG, and mammograms offers a huge volume of imaging data of large sizes. Radiologists and doctors do an excellent job of manually analysing and finding abnormalities. However, many rare and undiscovered diseases can make diagnosis a challenge. To help in such situations, machine learning can be used to recognize disease patterns from the large data sets amassed over the years. There are also pre-built classification libraries that have already analysed millions of pieces of labelled image data. These can assist doctors and healthcare professionals to diagnose patients correctly without the need for healthcare training.

Healthcare systems can also collect data from large technological companies such as Google and Apple. With the latest trend in wearable devices, patients want to collect as much data as possible regarding their health. These wearable devices can have additional attachments for diabetes or cancer patients to record additional data to promote their health and wellness. Google and Apple provide developer kits that can allow health systems to tap into these data stores and keep doctors connected with their patients. The combination of data from wearable devices and existing patient health records can provide additional insight and personalized healthcare solutions for a patient. Doctors can better manage patient conditions with wearable sensors, track conditions, and offer individualized treatment.

Health systems can also purchase a pre-built commercial platform that is ready to use and more user-friendly compared to an open-source custom-built solution for healthcare system. A powerful and well-known platform is IBM Watson, which is commercial software that is utilized for exchanging and investigating data between various hospitals, providers, and researchers. This commercial platform has the ability to extract maximum information from minimal input. Healthcare systems can easily set up these commercial out-of-the-box solutions and have continuous support from the company to troubleshoot any issues. These platforms are also validated and regulated for commercial release. Healthcare systems will have to pay the price if they choose to take this route but will have the benefits of a successful working analytics product.

The purpose of healthcare systems when adopting big data solutions is to make the patients' lives easier and healthcare more efficient. To achieve this, the health system will have the option to purchase a pre-built commercial solution where data can be stored and analysed. In order to get actionable insights for the benefits of the healthcare system and patients' health, monitoring starts from scratch and creates a custom-built solution that can be built and improved upon for years to come.

The custom-built solution will have to solve the volume, velocity, variety, and veracity problems the health system is facing. Starting with volume, the healthcare system will have to look into a storage solution. Many organizations prefer to keep data storage on premises to keep control over the data and maximize the uptime. However, on-site server networks can be costly to scale and retain over a period, especially as data grows at an exponential rate. Reductions in cost and enhancing reliability make cloud-based storage an appealing solution for data. Healthcare companies ideally should have a hybrid solution, which is the most flexible and usable approach for multiple data access and storage needs.

Once the storage systems are up and running, the foundational layer of data analytics is laid down. The next steps would be compiling the variety of data sources to be analysed. Healthcare data can come in many ways and there are multiple ways to store and collect the data, such as NoSQL database, mongoDB, or couchDB. Healthcare systems can also have the opportunity to deploy application programming interfaces (APIs) to transmit data, especially when data is housed in different locations dependent on where testing is completed. Once the data is compiled, we can use multiple choices of big data platforms and tools to begin performing big data analytics. The popular favourites are Hadoop and MapReduce to parse through large amounts of data and deliver queries and reports [13].

Big data algorithms can be applied in real-time for data pre-processing purposes. Healthcare systems should be able to develop a data pipeline to store large data collected from imaging such as MRI and CT scans in a format suitable for big data platforms such as Spark. This data conversion process saves time for downstream data cleaning and allows health systems to analyse large amounts of imaging data for quicker analysis. Other machine learning algorithms can be applied at this stage to parse through unstructured data to find patterns and anomalies in patients' health data. These algorithms can be trained and deployed to prevent disastrous disease outcomes for patients and assist doctors to deliver better care.

Using a custom-built solution for healthcare and deploying multiple open-source products come with a security risk. Healthcare systems will have to invest against malicious attacks intended to steal patients' data. Security systems will have to be fortified and follow the HIPAA security rules to store, transmit, and authenticate data and control access for specific individuals. Common security measures like using up-to-date antivirus software, firewalls, encrypting sensitive data, and multifactor authentication will have to be employed. Proper architectural solutions will be recommended, especially when handling patients' sensitive data, such as social security numbers, credit card information from billing, and test data. Such data will have to be encrypted and data access restricted to a limited number of employees. Machine learning algorithms can also be deployed to detect any network anomalies and alert hospital systems if the network is facing any type of cyber security attack with countermeasures deployed to prevent any access to patients' data.

1.1.2 Challenges in Healthcare

The biggest challenge in healthcare systems is conquering a variety of data and developing a pipeline to analyse multiple data types simultaneously. This allows insight to be gained into whether data is, for example, a patient's health, billing, or clinical testing data. The opportunities are endless for healthcare systems and the hurdle most of them encounter is where to begin. There is a large cost of investment required to analyse large quantities of data. The management of most healthcare systems would not invest in such areas since there is very little liquid cash flow available. The fear is that failed investments would bring loss to the organization, which would be catastrophic, and a larger financial challenge to recover from.

Resources are another hurdle healthcare systems will face. Full-scale data analytics for healthcare will need input from multiple sources including doctors, scientists, software engineers, and data scientists; all are costly to employ. These resources will have to spend time planning, building, validating, and ensuring the systems in place are working and delivering the correct information. The output of these analytics will need to be verified to ensure correct information is being relayed to the patients. The wrong treatment or options would be a costly error for patients and would be a blowback for the healthcare analytics project as a whole.

Big data is built around open-source projects such as Hadoop and Apache Spark. These open-source projects are much cheaper than the out-of-the-box solutions IBM Watson provides but these solutions come at a risk. There is not much technical support and a great many programming skills are required to successfully implement these tools. There is also a lack of security for open-source frameworks that is a risk healthcare systems must assess before they adopt such technologies. Healthcare systems are recommended to reinforce their network security to prevent a data breach.

The goal of big data in healthcare is to provide better treatment and a personalized diagnosis for patients. Healthcare systems can reduce costs through frequent testing, which reduces incorrect diagnosis and allows patients to get the best possible treatment at a faster rate. The investment for big data will reduce costs in the long run and keep healthcare operations running more efficiently than ever. Both patients and healthcare systems will reap the rewards of big data. One of the largest possible benefits of big data in healthcare is precision medicine – treatments and drugs that are specifically tailored to the patient's genome.

1.1.3 SEQUENCING GENOMIC DATA

The first human genome was sequenced in 2003 at a cost of $2.7 billion, called the Human Genome Project. It aimed to discover the base pairs that make up human DNA and was one of the novel scientific research findings. This massive achievement opened the door to greater understanding of human genes that could lead to tremendous advancements in the treatment of diseases. In future, it is possible that medicine will be "personalized" to a patient's individual genetic profile. In 2006, just three years after the first human genome was sequenced, Illumina sequenced a genome on its first machine for $300,000. Today, the cost has fallen to $1000, with the expectation that sequencing could be done for $100 in just a few years.

After an Illumina customer runs a sequencing task on one of their machines, the results are available on BaseSpace, Illumina's web interface for the sequencing data. BaseSpace allows its users to process, analyse, and store the genomes that they run and takes care of all the data security. A sequenced human genome is about 100 GB in size and it takes 24 hours for the instrument to sequence it. The computational aspect takes about 10 hours on a high-end cluster [13]. Today, BaseSpace's 90,000 users are generating a terabyte of data every single day, so a scalable platform is necessary. Illumina's data certainly qualifies as big data:

Volume: A terabyte of new data is being created every day with petabytes already in storage.

Velocity: The growth in sequencing data has been exponential over the past decade as prices drop.

Variety: The complete data is genetic data, but there is sequencing done for many different types of species and there is variation within species.

Illumina has been with AWS from the very beginning of their company. It became clear to them early on that sequencing genes on the scale and for the cost that they envisioned was going to take a massive amount of storage and computation. Advancements in computing power, storage, and the drop in price of computer hardware in the mid-2000s gave rise to cloud computing. Illumina probably would not have been able to develop as rapidly as it has, and sequencing pricing could not have fallen as sharply without the advent of the cloud. Outsourcing the "racking and stacking" of computer servers to a cloud provider enabled Illumina to focus on their key business: sequencing instruments and reagents. Illumina selected AWS for a variety of reasons, a few of which are mentioned below [13]:

- *Expertise*: As a cloud computing company, Amazon has the expertise in building the physical hardware and software that make up a platform. Using AWS's platform allowed Illumina to focus on developing cutting-edge gene sequencing machines, not computer infrastructure.

- *Scalability*: Illumina avoided the huge upfront costs of scaling up a network to meet current and projected needs, but at the same time, they were never constrained. They were able to deliver a consistent experience to their customers, even at times of peak demand.

- *Global Footprint*: AWS's global network of hubs gave Illumina the ability to move into international markets without building the back-end computer infrastructure.
- *Regulations*: AWS's physical hubs and security policies allowed Illumina to comply with privacy laws involving the handling of genetic data that can vary from country to country. Many countries insist that the genetic data of their citizens does not leave their borders.

Illumina uses many AWS services, but their main costs are elastic compute cloud (EC2) and simple storage solution (S3). The information created by their instruments is stored in S3 (10 petabytes as of 2016, much more than that in 2020). Computation is done with clusters on EC2. In 2014, in an effort to reduce its AWS costs, Illumina changed the composition of its EC2 clusters from all "on demand" to a mix of "on demand" and "spot instances". "Spot instances" enabled Illumina to bid on additional computational power at discounts of up to 90%. Overall, Illumina was able to decrease its monthly EC2 costs from $400,000 to just over $100,000. Another $90,000 was saved by switching some of its longer-term storage to S3 "infrequent access" [16].

Illumina's instruments are IoT enabled, and they use AWS Redshift as their data warehouse to manage the massive amounts of log data that these machines generate. As of 2016, Illumina's instruments were creating 270 billion data points per year [13]. Redshift allows them to sort through this massive amount of data to detect machines that may be faulty or require servicing. It also allows them to predict usage of reagents on a global, regional, and even individual customer scale. Illumina's success is the result of their brilliant scientists and engineers designing the world's best gene sequencing machines, coupled with the enormous advancements in computation that have developed over the last 15 years and are now so easily accessed over the cloud. AWS provides the platform on which BaseSpace, Illumina's software component, runs. This partnership allows Illumina to focus on gene sequencing and not computer infrastructure. Today, Illumina has a near monopoly on the sequencing instrument business due to the breakthroughs in their devices as well as their leaders' abilities to answer their configuration management issues.

1.1.4 Deep Learning Applied to Genomic Data

The proliferation of sequencing machines has led to an explosion in genetic data. The interpretation of this data is one of the great questions of our time and the most challenging big data AI problem to date. While the sequencing side of genomics is primarily a device-driven business that is narrow and largely controlled by one company, the interpretation of the data is wide open with a variety of different avenues of research from both business and academia. However, the biggest challenge was to interpret the structure, function, and meaning of the genetic information itself. Biologist Eric Lander summarized the situation in seven words: "Genome, Bought the book. Hard to read" [12]. This quote really sums up the reality of genomics today. We are generating massive amounts of sequencing data but are only at the very tip of the iceberg in our ability to use the information. It should come as no surprise that we cannot intuitively read the code of C, A, T, and G that makes up the instructions for all life. However, strides are being made all the time and we have gleaned some useful information for genomics.

For instance, "Spinal muscular atrophy (SMA), which is the leading genetic cause of infant mortality in North America, results in the baby's genome missing the SMN1 gene, or contains a damaged version of it, resulting in deficient production of the survival motor neuron" [12]. In this case, we know exactly the gene and the corresponding disease. Today, we can test for the disease and researchers are now experimenting with various new treatments.

Predicting phenotype (traits and disease risks) from the genotype (underlying genetics) is the overarching goal of genomics and is somewhat like a typical machine learning problem, but also quite different. There is a target variable (a particular trait or disease), and genome features are used to make accurate predictions. "The human genome has 20,000 protein coding genes, and 25,000 non-protein coding genes. Some genes are crucial for life, and some can be deleted in their entirety without apparent harm" [13]. To say that the genome is a noisy data set is an understatement. In addition, many of the factors that determine overall phenotype (cancer diagnoses, for example) are not contained in the genotype at all (perhaps the patient smoked). A better method for modelling disease risk is to predict intermediate cell variables, also known as molecular phenotype, first and then this variable in turn can predict overall phenotype. In other words, genomics models that predict intermediate cell variables suffer from less bias than models that try to predict overall phenotype. Think about the example of identical twins: their genomes are the same, yet these two people may experience vastly different health outcomes based on environmental factors.

Most of the tremendous machine learning advances we have seen in the past decade are tasks that are actually very easy for a human – identifying an object, reading a tweet, or picking up a small object. These tasks are easy for us to frame as machine learning problems, because they are things we intuitively understand, such as seeing. We are adapted to these very "human" tasks. However, the machine learning problems posed in genomics are definitely not "human" problems. We do not intuitively understand DNA. In fact, unravelling the code embedded in gigabytes-worth of information stored on microscopic protein chains is a uniquely "un-human" task. Our brains cannot process that amount of information and none of our senses is useful in observation. The challenge posed by genomics is harder than most machine learning tasks because it is a task that we cannot do ourselves, and the fact that it is a task that humans cannot do is precisely why it is so important that we find a way to frame it as a machine learning problem. The people who best understand molecular biology and genetics, as in the people who can best ask the questions that machine learning might solve, usually lack the skills in data science to perform the modelling tasks.

To bridge this gap, companies are creating cross-functional teams of data scientists working with molecular biologists and geneticists. One such company is Deep Genomics. Deep Genomics is a small Canadian start-up using geneticists, biologists, chemists, toxicologists, clinicians, and computer scientists to leverage AI in the discovery of new drugs. They say on their website, "The future of drug development will rely on artificial intelligence, because biology is too complex for humans to understand." To date, the company has discovered a promising compound to treat Wilson's disease using their AI platform – a first for AI in medicine [17, 18].

Another solution is coming from the cloud. All the major cloud players now offer genomics platforms to help automate or at least streamline the data science portion of analysing the genome. Microsoft, for instance, offers Microsoft Genomics on Azure. "Microsoft is deepening its commitment to genomics with the fourth iteration of the open-source Genome Analysis Toolkit (GATK4). The toolkit update is designed to optimize performance for researchers as they strengthen data pipelines and power successful genomics analyses, so they can reduce the risk of noise or artifacts within the data set and extract more insight from the genome" [14]. Microsoft is positioning this service to use the impressive power of the Azure cloud in conjunction with a streamlined front-end in the GATK4 toolkit, so that researchers can conduct machine learning without building their own data pipelines and algorithms from scratch. Unsurprisingly, AWS and Google cloud also offer very similar genomics packages on their clouds. While not having the monopoly that it has in sequencing, Illumina's BaseSpace also offers analysis capabilities and provides a full end-to-end solution for many of its instrument users.

So far, only the challenges in interpreting the genome via machine learning have been discussed, but it should be stressed that the only way we are ever going to understand the human genome is through the combination of human intelligence and machine learning. The genome is simply too large to be analysed without a computer doing the heavy lifting. As stated earlier, predicting phenotype from genotype is essentially a supervised learning task. In the mid and early 2000s, Support Vector Machines were the most popular algorithm for genomics classification. Nowadays, the most common algorithms are deep neural networks such as tensorflow. Tensorflow was originally designed by Google to improve its website services such as search or Gmail. However, it has proven to be a highly accurate classifier in many other domains such as genomics. Many types of neural networks can be used to model genomics data, but the convolution neural network (CNN) has proven to be especially accurate [17]. As with all model building, more flexible models such as CNN should be compared with more simple models such as support vectors or boosted trees. One of the main drawbacks of neural networks is their lack of interpretability. In many genomic applications, researchers are more interested in gleaning insights from the genome than in high classification accuracy. This problem is inherent to neural networks (NN) across all domains, and bringing interpretability to the model is the forefront of NN. One such variant is called Deeplift, which seeks to give the user feedback on feature importance. However, Deeplift is a work in progress and still not widely adopted [17].

Interpretation of the sequenced genome might be harder than sequencing the genome itself. This domain is a wide-open opportunity for research, and the field of genomics is still in its infancy. Because the data is so vast and unobservable by humans, machine learning must play a role in the unravelling of the human genome. However, this task comes with substantial configuration management issues.

1.1.5 GENOMIC DATA AND MODERN HEALTHCARE

There are applications today where normal people are benefiting from the breakthroughs in genomics over the past 20 years. Two such examples are Veritas, a personal whole-genome sequencing service. Another is a partnership between Quest Diagnostics and IBM. Veritas is a company that offers whole-genome sequencing and interpretation for its customers. While companies like Ancestry.com perform genotyping, a surface-level analysis of the origins, Veritas digs deep and attempts to interpret your whole genome. They can offer not only basic information on your genotype, but also screening for hundreds of diseases, your genetic response to certain medications, if you are a carrier for genetic diseases, and lifestyle attributes like athleticism or metabolism. They provide the customer with all the information that they can currently glean from sequencing a full genome. What is truly amazing is that they do this for a price of only $599, with the goal that they will someday be able to offer it for $100. Veritas uses Microsoft Genomics on Azure and Illumina sequencers.

Quest Diagnostics and IBM, two companies we have all heard of, created a partnership in 2016 to help diagnose and treat cancer. The service works like this: a physician sends a solid tumour biopsy to Quest Diagnostics. A Quest pathologist prepares the tumour sample and sequences the sample on an Illumina machine. The results are uploaded to IBM Watson, which analyses the raw data to identify the particular mutations that have occurred in the tumour. Watson then searches a massive corpus of medical literature to identify the best therapeutic options for that particular tumour. Memorial Sloan Kettering Cancer Center, a renowned cancer research centre, is one of the sources feeding Watson. The tumour data and recommendation are then sent back to the Quest pathologist and ultimately the physician and patient [14]. This is a practical example of the type of precision medicine promised by the Human Genome Project 20 years ago bringing benefits to everyday people.

1.2 BACKGROUND AND RISE OF INTERNET OF THINGS

The Internet of Things (IoT) is a network of physical devices having the ability to sense and collect data from the user or the world around, later utilizing the data collected in various aspects. The web of the IoT currently contains almost 7 billion devices and is expected to increase to 25 billion by 2025. IoT is a new field of research and its health-related possibilities are currently unacknowledged. One the most important organs of the human body is the heart. Any problem with the heart may lead to various serious outcomes. According to the World Health Organization (WHO), more than 18 million deaths around the world every year (that is around 31% deaths of total deaths worldwide) are due to cardiovascular diseases (CVDs), making it the number one cause of death worldwide.

With the help of the IoT, monitoring and collection of data is very easy and feasible as data can be collected from anyone anywhere, and the sharing of the data is also very cost-effective because it is stored on the cloud. Various data related to CVDs can be collected through a smartphone application or wearable device. Through this data, the device can trigger an alert if the user is exhibiting the same heart pulses as those of patients with CVDs whose data is gathered in the cloud through the IoT network of connected devices. Since there is an increase in the number of elderly people and in the number of patients with chronic diseases, the traditional health support systems are demanding a change and it is very encouraging. However, there is a lack of utility as the required facilities are only available in hospitals and are not readily usable by elderly and disabled patients, with the result that they do not satisfy the needs posed by critical conditions. This calls for an advanced thought – pervasive healthcare, providing healthcare support to everyone, everywhere, around the clock, addressing all the challenges effectively. Plenty of pervasive healthcare applications have been introduced in the recent past. Uniyal reviewed existing pervasive healthcare applications, focusing on different living conditions of humans such as elderly people living alone, disabled care support, or support for diseases such as Parkinson's, heart diseases, and diabetes [19]. All the mentioned pervasive healthcare systems take care of various factors such as real-time monitoring, running incidence detection algorithms, emergency intervention, and patients' self-management.

1.2.1   IoT in Real-Time Healthcare Applications

1.2.1.1   Wearable Front-End Device

A wearable front-end device is used to collect ECG data from different people and an AI-based algorithm is used to identify failures. This front-end device is a system on a chip (SoC) and is integrated in smart watches. With the help of Bluetooth, the front-end device sends data to a smartphone application.

1.2.1.2   Smartphone Application

An application connected to the SoC device mentioned above is used to display the data, and has a function to show ECG signals to the user when the alert has been triggered by any irregularities. The safety features include a function to suggest various measures to reduce the irregularities, for instance yoga. Finally, the DATA AND ANALYSIS function collects the data from the SoC and transfers it to a cloud-based storage system where all the algorithms are present and analysis of data occurs. Later, the analysed data is sent to the user with a failure percentage so they may consult doctor if needed.

1.2.1.3   Cloud and Algorithms

With the advancement in cloud storage, storing and accessing data has become easy. This process mainly includes the cleaning, encryption, and analysis of data with algorithms and returning the required results. First of all, when the data being sent to the cloud reveals no useful information, cleaning and compression of data takes place, after which the encryption of data helps to secure the data from leakage or exchange by giving it unique key IDs. This data passes through two segments: through an algorithm and through doctors. Doctors can diagnose these ECG signals from patients using a web application. After the data has been analysed by the doctor and algorithm, it returns to the mobile application and then to the user."

1.2.1.4   How Does It Work in Real-Time?

The network of devices used as a platform for this study includes a wearable device, which analyses the ECG signals in real-time to identify if there is any disturbance in the heartbeat. With the help of a smartphone, the application compares the user's heart data with the data present in the cloud and triggers the alarm if something seems to be wrong. By using the data collected from the cloud and an AI-based algorithm, further enhancements and analysis of the results can be done.

1.3   SUMMARY

This chapter reviews the IoT and big data platform, integrating healthcare systems, cloud databases, ECG hardware, AI, and machine learning software, expecting to improve the health status of human beings. Placement of electrodes on one's own skin and uploading the ECG signals are possible today with the advent of IoT, reducing the possibilities of any mishandling. The involvement of IoT and big data in healthcare applications makes it possible for patients to receive critical care before they reach the hospital and take the required precautions to stop or reduce the effects of the irregularity. In this chapter, we presented an overview of big data in the healthcare system and the positive impact it can have in the lives of patients. We analysed a case study of an IoT-based heart disease monitoring system for a pervasive healthcare service to illustrate the benefits of advanced technical support in healthcare contexts.

REFERENCES

[1] Sarraf, S. and Mehdi, O., "Big data spark solution for functional magnetic resonance imaging," *arXiv.org*, March 23, 2016. https://arxiv.org/abs/1603.07064.

[2] Saheb, T. and Izadi, L., "The paradigm of IoT big data analytics in the healthcare industry: a review of scientific literature and mapping of research trend," *Telematics and Informatics*, vol. 41, pp. 70–85, 2019. doi: 10.1016/j.tele.2019.03.005.

[3] Kalejahi, B. K., Saeed, M., Yariyeva, A., Dawda, N., Maharramov, U, and Ali, F., "Big data security issues and challenges in healthcare," *arXiv.org*, December 9, 2019. https://arxiv.org/abs/1912.03848.

[4] Ishwarappa, J. A., "A brief introduction on big data 5Vs characteristics and hadoop technology," *Procedia Computer Science*, vol. 48, pp. 319–324, 2015. https://doi.org/10.1016/j.procs.2015.04.188, ISSN: 1877-0509.

[5] Raghupathi, W. and Raghupathi, V., "Big data analytics in healthcare: promise and potential," *Health Information Science and Systems*, vol. 2, no. 1, p. 3, July 2014. https://doi.org/10.1186/2047-2501-2-3.

[6] Kanehisa, M., Goto, S., Hattori, M., Aoki-Kinoshita, K. F., Itoh, M., Kawashima, S., Katayama, T., Araki, M., and Hirakawa, M., "From genomics to chemical genomics: new developments in KEGG," *Nucleic Acids Research*, vol. 34, no. 1, pp. D354–D357, January 1, 2006. https://doi.org/10.1093/nar/gkj102.

[7] Sethy, R. and Panda, M., "Big data analysis using Hadoop: a survey," *International Journal of Advance Research in Computer Science and Software Engineering*, vol. 5, no.7, pp. 1153–1157, 2015.

[8] Sultana, A., "Using Hadoop to support big data analysis: design and performance characteristics," *Culminating Projects in Information Assurance*, MSc thesis, Dept. of Information Systems, St. Cloud State Uty., Minnesota, 2015.

[9] Pol, U., "Big data and Hadoop technology solutions with Cloudera manager," *International Journal of Advanced Research in Computer Science and Software Engineering*, vol. 4, pp. 1028–1034, 2014.

[10] Lee, J.-R., Khan, N., Yaqoob, I., Hashem, I. A. T., Inayat, Z., Mahmoud Ali, W. K., Alam, M., Shiraz, M., and Gani, A., "Big data: survey, technologies, opportunities, and challenges," *The Scientific World*, vol. 2014, p. 18, 2014.

[11] Oussous, A., Benjelloun, F.-Z., Ait Lahcen, A., and Belfkih, S., "Big data technologies: a survey," *Journal of King Saud University – Computer and Information Sciences*, vol. 30, no. 4, pp. 431–448, 2018. https://doi.org/10.1016/j.jksuci.2017.06.001, ISSN: 1319–1578.

[12] Leung, M. et al., "Machine learning in genomic medicine: a review of computational problems and data sets," *Proceedings of the IEEE*, vol. 104, no. 1, pp. 176–197, January 2016.

[13] Dickeron, A., "Illumina massively scales its DNA sequencing technologies using AWS," *Youtube*, 2016. Available: https://www.youtube.com/watch?v=MJdPUws_SaQ, Accessed: Dec.12, 2019.

[14] Miller, G., *"Microsoft unveils genomics innovations and new partners at ASHG 2018,"* https://cloudblogs.microsoft.com/industry-blog/health/2018/10/15/microsoft-unveils-genomics-innovation-and-new-partners-at-ashg-2018/, Accessed: Dec. 2, 2019.

[15] Dash, S., Shakyawar, S. K., Sharma, M., and Kaushik, S., "Big data in healthcare: management, analysis and future prospects," *Journal of Big Data*, vol. 6, no. 1, pp.1–25, 2019. https://doi.org/10.1186/s40537-019-0217-0.

[16] Amazon Web Services, "*Illumina case study*," https://aws.amazon.com/solutions/case-studies/illumina/ Accessed: Nov. 25, 2019.

[17] Zou, J., "A primer on deep learning in genomics," *Nature Genetics*, vol 51, pp. 12–18, January 2019.

[18] Lawrence, S., "Quest, IBM bring artificial intelligence to genomic tumor data," *FierceBiotech*, October 17, 2016. https://www.fiercebiotech.com/medical-devices/quest-ibm-bring-artificial-intelligence-to-genomic-tumor-data.

[19] Uniyal, D. and Raychoudhury, V., "Pervasive healthcare – a comprehensive survey of tools and techniques," *Computing Research Repository*, vol. 1411, no. 1821, ArXiv abs/1411.1821, 2014.

2 Securing IoT with Blockchain

Challenges, Applications, and Techniques

Vidushi Agarwal and Sujata Pal
Indian Institute of Technology Ropar

2.1 INTRODUCTION

The Internet of Things (IoT), which is evolving rapidly in the industrial and academic areas, is a promising technology aiming to revamp human life completely. The IoT is a heterogeneous network of interconnected devices with the capabilities to actuate, sense, store, and communicate, thereby connecting real-life to the digital world. The IoT can help in enhancing the quality of products and gaining profits through autonomous operations and increased product throughput. In today's era, an IoT device has varying functionalities ranging from a wearable device to complete industrial automation. The IoT has the power to transform a home into a smart home, energy grids to smart energy grids, a city to smart city, and much more.

However, the data collected by IoT devices is not entirely secure and tamper-proof, which may pose a threat to the privacy and confidentiality of users [1]. Moreover, as the number of IoT devices is increasing, the traditional cryptographic methods are no longer able to perpetuate data integrity against the threats and risks they are exposed to. Basically, IoT systems are made up of four layers, as shown in Figure 2.1.

- The first layer is made up of the battery-controlled, resource-constrained, small devices with attached actuators and sensors. These devices sense and collect data, perform lightweight processing on it if required, and then communicate the data to the network.
- The second layer, the IoT gateway, is a powerful processing device, which aggregates data from various devices and delivers this data further to the cloud services.
- The next layer is the cloud, which is used as a storage server for all the collected sensor data. The data can be analysed as well as processed in these cloud servers for decision-making.

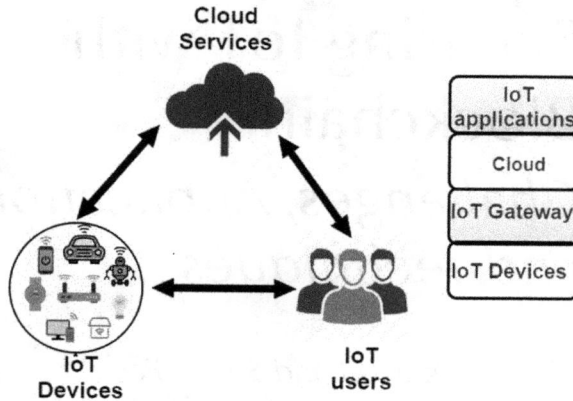

FIGURE 2.1   Traditional IoT services.

- The fourth layer is the platform layer where the raw data is used for business services and user applications by applying data analytics, statistical methods, machine learning, and other complex algorithms to extract corrective results from the data.

IoT systems are dependent on cloud services for data processing and storage, but these centralized cloud servers are also susceptible to intruder attacks, single point failures, data tampering, and cyber-attacks. This hinders the adoption of IoT in a large-scale scenario. Most of the existing methods of security rely on third parties to handle users' data, which can lead to misuse and unauthorized sharing of private data.

Being a tamper-proof, immutable and distributed ledger-based technology, blockchain is the key to solving the security issues of the IoT. Blockchain can provide the security needed for the IoT with its distributed nature and features like secure authentication, accountable data sharing, and traceable auditing. Blockchain solves the authentication problem of the IoT by managing a public–private key and using "digital signatures" to authorize and authenticate clients. Blockchain is secure since it stores transactions of data in the form of a distributed ledger, making chains of blocks validated by using consensus methods.

In this chapter, we study systematically the current research techniques to understand how blockchain can be used in IoT systems. We initially study the security risks of the IoT and how these risks can be alleviated using blockchain. The fundamentals of blockchain are then explained, along with the advantages and applications of blockchain-based IoT infrastructures. Finally, we analyse and categorize the existing integration methodologies based on the approach used for integration of blockchain with the IoT. Therefore, the major contributions of this chapter are highlighted as follows:

- A summary of the security issues in IoT
- A summary of the fundamentals of blockchain technology and the types of blockchain

- An overview of the integration of blockchain and IoT and how blockchain can address the issues in IoT
- Real-life applications of blockchain-based IoT architectures
- Analysis of the existing techniques used to integrate blockchain and IoT

The remainder of the chapter is organized as follows. Section 2.2 discusses the security issues of the current IoT scenario. Section 2.3 introduces the fundamentals and types of blockchain. Section 2.4 presents how the integration of blockchain can address the issues of the IoT. Various real-life applications of the integration of the IoT and blockchain are discussed in Section 2.5. In Section 2.6, we categorize the blockchain-based IoT architectures based on the approach used for their integration. Finally, Section 2.7 presents a brief conclusion and recommendation on the future work that can be done in this area.

2.2   SECURITY ISSUES OF IoT

Since the number of IoT devices is increasing each day and they are often deployed in hostile, unattended, and unfavourable conditions, securing them becomes a colossal challenge. Several incidents and attacks involving IoT devices have been devised in the past few years, which makes it difficult to trust them. In 2018, it was discovered that smart assistants like Alexa and Google home were exploited by hackers to snoop on users without their knowledge. A smart refrigerator sent a large number of email spam messages in 2014 without the owners being aware. The large quantities of data generated by IoT devices, open wireless channels, and the complexity of IoT systems further adds to their security risks. The following subsections introduce some of the typical security issues in IoT networks.

2.2.1   IoT MALWARE

IoT devices are exposed to various security problems, which can compromise the services provided by them. One such risk that makes them vulnerable is the attack by malware, which is created by hackers for the purpose of stealing data, damaging devices, or simply causing a mess. Malware is malicious software, which can be of many types, such as viruses, worms, ransomware, and Trojans. Wang et al. [2] categorized IoT malware into two types based on how the IoT devices are corrupted: one exploits IoT devices through their unfixed or imprudent vulnerabilities, while the other uses brute force methods to infect devices. The most common IoT vulnerabilities are buffer overflow, poorly implemented encryption, unencrypted services, and denial-of-service attacks. One of the reasons that makes brute force attacks successful is the use of weak passwords.

2.2.2   DEVICE UPDATES MANAGEMENT

Although when purchased, the IoT devices could contain the latest security software, it is not possible to avoid any new risks or attacks which may arise later. Therefore, keeping IoT devices updated with the latest software becomes a necessity instead of an option. However, IoT updates are still not delivered as efficiently as those delivered to smartphones and computers as the manufacturers of IoT devices pay less attention to the security risks. The updates should be delivered automatically because the customers feel it is not their duty to carry out updates and we cannot expect users to stay on top of every software update. For the IoT devices that are deployed in remote areas or are difficult to access, over-the-air updates are an option, but they can also pave a path for the hackers to use malicious software for updating IoT devices.

2.2.3 Manufacturing Defects

IoT device manufacturers are more eager to release their products in the market than to test for possible security leaks. This is among one of the major security issues of the IoT. Without any standard security mechanism, IoT manufacturers will keep producing devices that are vulnerable to attacks. For example, a smart refrigerator can compromise a user's login credentials for their Gmail ID, and most fitness bands available today remain visible to be paired by Bluetooth even when they have already been paired by the user. Moreover, these resource-constrained devices cannot support heavy methods for security because complex encryption and decryption methods cannot be performed fast enough by them to transfer data in real-time securely.

2.2.4 Security of Massively Generated Data

With the increase in IoT devices each day, the data generated by them is also increasing. Devices like smart thermostats, smart TVs, lighting systems, and speakers constantly produce data leading to a problem of processing, transmitting, and storing it securely. This data should be kept encrypted or anonymous before it is stored or sent to the unsecure cloud servers so that the personal information of users cannot be revealed. Disposing of cached or unwanted data is also a security challenge so that it does not fall into the wrong hands. Moreover, data integrity should also be maintained by using digital signatures or checksums. It is a common practice nowadays that the service providers sell or share this data with other companies without the permission of users, thus violating their privacy and trust. Therefore, ensuring the privacy of users' crucial data has become a serious concern to protect them from hackers.

2.2.5 Authorization and Authentication Issues

Authentication is the process of identifying whether the user is the same person/ entity they are claiming to be. Authorization means checking whether the user has the required permission for the data or resources they want to access. In machine-to-machine communications, authentication and authorization should be provided by the use of cryptographic techniques [3]. Many IoT devices fail to authenticate other devices properly because the basic means of authentication is the use of passwords and users can use weak passwords and even default passwords, which are easily predictable. The authorization mechanisms available currently are mostly centralized and do not provide efficient, effective, manageable, and scalable mechanisms to control and verify the access permissions of the users accurately. The increasing number of IoT users and data produced by them makes it even more necessary to devise scalable and manageable authorization techniques.

2.2.6 BOTNET ATTACKS

A botnet is a cluster of devices connected through the Internet whose security has been compromised by an attacker. Hackers create botnets by using malware to corrupt Internet-connected devices and then controlling them through a server. If one device on a network has been compromised, all the other connected devices pose a threat of infection. One such example which affected IoT devices is the Mirai botnet (2016) [4]. The Mirai botnet led to a denial-of-service attack with 620 Gbps of traffic led by a group of IoT devices, security cameras, and routers. A denial-of-service attack leads to the unavailability of network or machine resources for legitimate users due to a malicious actor. A perpetrator overloads the targeted machine or network with traffic until no more legitimate requests can be satisfied or processed. At almost the same time, another denial-of-service attack targeted the French webhost, which used the Mirai malware with a peak rate of 1.1 Tbps traffic. Relying on the same principle, many other botnets have been discovered since. The Hajime botnet (2016), relying on distributed communications unlike Mirai, and the BrickerBot (2017), an IoT botnet based on BusyBox, are two such examples. To avoid the common vulnerabilities, it is necessary to adopt some standard security mechanisms to ensure that IoT devices are not used as zombies. Proper intrusion detection systems should be installed because IoT software codes might become obsolete when not updated for long periods of time.

2.3 INTRODUCTION TO BLOCKCHAIN

A blockchain is basically a record of blocks, which contain digital information, secured via "digital signatures". It is a distributed ledger used to save information about transactions between two parties in an incorruptible way. The distributed ledger technology offers an advanced way of identity management through public and private keys instead of passwords, which could be cracked easily. Each block contains a hash of the previous block, which means that if an intruder wants to modify one block, he will have to modify all the preceding blocks, thus making it permanent. Blockchains offer a secure and transparent way to exchange information as every node in the blockchain has a local copy of the blockchain and any user can verify the identities themselves.

The transactions in blockchain are validated by certain nodes called "miners". The miners use the consensus mechanism as proof of work to validate the transactions. The transaction is valid only if the hash of a block with reference to the hash of its preceding block is correct. The validated transaction is then combined with other transactions, and once the consensus is reached by a majority of nodes, the new block is added to the existing blockchain network. The existing local copies of the blockchain are then updated on all the nodes with the new block. When two parties wish to exchange services or information, blockchain uses smart contracts similar to paper contracts for stating the terms and conditions of the deal. Smart contracts are digital, self-executable codes that run automatically when some predefined conditions are met. Smart contracts stored on the blockchain are helpful to set up trustworthy relationships among parties without the need for any third party. Ethereum is a public, open source, decentralized blockchain-based platform featuring the functionality of smart contracts. Smart contracts have applications in fields like healthcare for health insurance, industry for financial agreements, and real estate for documents of property owners. There are mainly three kinds of blockchain, namely public blockchain, private blockchain, and consortium blockchain.

2.3.1 PUBLIC BLOCKCHAIN

Public blockchains are open to everyone for accessing, reading, and writing to the ledger. Anyone can join the blockchain network and verify the transactions to be added to the blockchain. Decision-making in such an open and decentralized system is managed by consensus methods such as proof of stake and proof of work. These blockchains are completely decentralized without the intervention of any central authority to modify the ledger or smart contracts, which makes them reliable against any single point of failure. Also called permissionless blockchains, public blockchains are used mainly for exchanging and mining cryptocurrencies such as Bitcoin and Ethereum.

2.3.2 PRIVATE BLOCKCHAIN

Private blockchains, also known as permissioned blockchains, have restricted access on who can participate in the network. Only the users chosen by the respective authority or network administrator have permission to read or write in the ledger. Such blockchains are used mainly by centralized organizations who want to store their transactions privately in a closed network. In private blockchains, everyone knows the identity of users but transactions can only be viewed by those with the appropriate permission. Such blockchains are more efficient and have higher throughput due to the finite number of users. Some examples of permissioned blockchains are Hyperledger and Multichain.

2.3.3 CONSORTIUM BLOCKCHAIN

Consortium blockchains are a hybrid of public and private blockchains where some nodes are in charge of the consensus mechanisms, while other nodes have access to the transactions. Unlike private blockchains, these blockchains are governed by more than one organization instead of a single entity. In these blockchains, transactions can be made and blockchains can be edited/reviewed only by chosen members of the consortium (federation). This approach has the benefits of a private blockchain and is very efficient, since all the parties involved work together for the overall benefit of the network. These blockchains are used by governments and central banks to supply chains. Some examples are Energy Web Foundation (EWF) and R3 blockchains.

2.4   BLOCKCHAIN AND IoT INTEGRATION: AN OVERVIEW

Since the IoT devices sense and collect users' personal data in applications like healthcare and smart homes, the protection of these devices should be a major research priority. Moreover, these devices have the capability to control the physical environment of users, so attacks on these devices could lead to major risks to human life. Blockchain can be used to resolve major security risks associated with the IoT by providing a secure method to share and store data in an incorruptible form. Data stored in a blockchain is traceable and can be backtracked to its source at any point of time. The number of IoT devices being deployed is increasing, so the trend has to shift from the use of a central cloud server to a more distributed kind of technology like blockchain for the security of data and IoT devices. The integration of blockchain and the IoT has the following advantages:

- **Authorization and Authentication Management:** Authentication of IoT devices is a major issue that can be solved by using the "digital signature" feature of blockchains. A blockchain system can identify each IoT device uniquely and also the data stored by each device. Hammi et al. [5] proposed a method to securely identify and authenticate IoT devices by using a decentralized blockchain-based method. ControlChain [6] is a blockchain-based architecture for authorization of IoT to maintain the confidentiality of stored data.
- **Scalability:** The massive growth of IoT devices is generating a large volume of data which is difficult to transmit and store securely in a centralized architecture. Therefore, a scalable and distributed structure like blockchain has become necessary for privacy and security of data. Moreover, using blockchain for data storage can prevent single points of failure and enhance fault tolerance. One such distributed scheme for data storage has been proposed by Li et al. [7] based on blockchains and certificateless cryptography.
- **Security and Reliability:** IoT devices are prone to intruders in ways that cannot even be predicted by manufacturers. The limited computational resources of IoT devices hinders the use of complex security techniques. To provide data security, blockchain can be employed, since it stores all the information using transactions, which are protected and validated through cryptographic encryption. Khan and Salah [8] present a survey on how blockchain can be the key to solve most of the security issues in IoT. Blockchain makes IoT data reliable because the data stored in the form of transactions is immutable and cannot be altered. Moreover, the participants can trace the data back and verify it at any point in time.
- **Autonomous Environments:** Blockchain technology allows IoT devices to interact automatically without the intervention of third parties or any central authority. Smart autonomous environments can be developed in IoT through the secure and distributed nature of blockchain-based architecture. Blockchain combined with smart contracts is a powerful technique which can convert IoT devices into decentralized autonomous corporations (DACs). These DACs are capable of interacting and making decisions independently, which makes this technology appealing to IoT developers.

- **Interoperability of IoT Devices:** IoT devices are supported by various types of agencies, such as multinational or independent bodies, organizations, and alliances, making them different in terms of architecture, communication technologies, and other services. Blockchain technology can be used to enhance the interoperability of heterogeneous IoT devices by processing their data and storing it into the common blockchain.

2.5   APPLICATIONS OF INTEGRATION

The innate characteristics of blockchain lead to its integration in several applications of IoT. Since blockchains can mitigate the various security risks of IoT as discussed in the previous section, it makes them an appropriate choice to enhance the privacy of current IoT applications. Some illustrative applications of blockchain are introduced in the following subsections.

2.5.1   SMART HOMES AND CITIES

A smart home uses smart devices connected through the Internet to manage home appliances, such as heating and lightning, remotely. IoT devices are used in smart homes to automate smart devices and connect them to each other as well as to the Internet. However, such devices have limited processing capabilities and are not entirely secure, posing a threat of cyber-attacks on the smart homes. Attacks on a smart home can lead to stealing users' private information or controlling access to theft control systems, electrical appliances, door locks, gas leakage detectors, and so on. Blockchain-enabled IoT devices provide high security to the users and immutable access to the smart devices without any leakage of information. One such architecture is proposed by Dorri et al. [9] which uses a high-resource device called a "miner" for handling all the communications and a local blockchain for monitoring the transactions. A smart city is a similar area which uses different IoT devices to collect information and analyse the data collected to manage the resources and assets of a city in an efficient manner. One such example is proposed by Biswas et al. [10] where blockchain is combined with smart cities to prevent digital disruption of private information.

2.5.2   HEALTHCARE

One of the major persisting socio-economic problems in today's scenario is healthcare. Due to the limited number of hospital resources for a large population of patients, the classical way of handling patients has become very difficult. The recent advancement of wearable devices and electronic health records in healthcare has promoted the usage of remote healthcare services, reducing the burden on hospitals' resources and providing doctors with live tracking of their patients' medical data along with automatic warning signals in case of any emergency. However, remote access to the medical data also raises some security and privacy issues. To prevent the data from getting into the wrong hands, the security of healthcare data and devices can be ensured by integrating blockchain into the healthcare network. Griggs et al. [11] proposed a blockchain system for healthcare in which the smart contract system maintains the real-time monitoring for patients as well as medical professionals by sending notifications to both of them. Zhang et al. [12] described how blockchain can be used in the field of healthcare. They also described the various challenges we face while using this technology in healthcare.

2.5.3 INTERNET OF VEHICLES

The integration of IoT in vehicles will result in the evolution of an era when smart communication between vehicles and between vehicle and road, vehicle and human, vehicle and sensors, vehicle and environment, and vehicle and everything will be established, such that all the vehicles will be connected to a single network. However, to handle the transmission and managerial aspects of the vehicles, a blockchain-enabled Internet of Vehicles (IoV) may be used to resolve problems like resource scheduling and broadcast collision-avoidance. Sharma [13] proposed a transmission model that is energy-efficient and can optimally handle the demands of a blockchain-enabled IoV by regulating the count of transactions via a clustering mechanism. Jiang et al. [14] explored how the vehicle network, when cascaded with blockchain technology, can result in a solution to various applications associated with distributed and secure storage.

2.5.4 SMART MANUFACTURING

With the rapid increase in technology, the advancement from automated manufacturing to smart manufacturing can be widely experienced by the manufacturing industry. During each and every phase of the production cycle, such as product designing, supply of raw materials, product manufacturing, product retailing, market supply and export, and product sales, substantial amounts of data are generated. Data analytics and information extraction cannot be directly applied to generated data as it is inconsistent and fragmented. At this point, blockchain comes to the rescue by handling interoperability and providing shared access to data in the manufacturing sector. Abeyratne and Monfared [15] reviewed the present scenario with this technology in the manufacturing sector. Various advantages and probable benefits of this technology in the manufacturing supply chain are described along with its future perception. BPIIoT [16] is a peer-to-peer blockchain-based platform that is decentralized in nature. Without the need of an intermediate trustee, this platform helps peers interact among one another.

2.5.5 SUPPLY CHAIN

Manufacturing a product may involve some parts being shipped from other countries. This may lead to some low-quality products seeping into the supply chain. Blockchain can be used to address this issue by using unique identifiers for every part of the supply chain. Since every node in the blockchain can store a copy of the ledger, every transaction is verifiable. All the copies will have to be changed simultaneously if anyone wants to tamper with the blockchain. Blockchain can help in improving supply chain management by recording the product's quantity and the parties supplying it, tracking all the details related to trades, verifying the quality and other standards for products, and sharing this information among different parties in the supply chain. Some benefits and innovations for blockchain-based supply infrastructure have been proposed by Saberi et al. [17]. Kshetri [18] studied the effects of blockchain on supply management in terms of flexibility, speed, quality, sustainability, cost, risk reduction, and dependability.

2.5.6 SMART ENERGY GRIDS

As technology is advancing and the IoT is being used in energy systems, the trend is shifting to smart energy grids. A smart grid is an energy grid which combines smart digital communication approaches with the traditional electric networks. A smart grid involves energy measures like smart applications and smart meters, and it produces power using energy-efficient resources like solar systems, wind turbines, renewable resources (e.g., bio-fuels), and even modern electric vehicles. IoTs can help smart grids by providing functionalities like real-time pricing, load forecasting, load monitoring, and demand response. Using renewable energy resources like solar panels, the energy consumers can also help in producing energy. Energy trading between such producers and consumers becomes difficult due to lack of trust and insecurity. Blockchain can help in energy trading between producers and customers using transactions. One such system based on smart contracts is proposed by Mylrea and Gourisetti [19] for energy trading based on predefined rules without the intervention of third parties. Another blockchain-based approach for transaction security is proposed by Aitzhan and Svetinovic [20] for decentralized energy trading via a proof-of-concept mechanism for securing transactions and anonymous negotiation of energy prices.

2.6 EXISTING RESEARCH ON BLOCKCHAIN-BASED IoT SECURITY

Integration of blockchain with IoT gateways and end devices can be achieved through several methods, depending upon the available resources and capabilities of IoT hardware. Figure 2.2 shows how IoT nodes can communicate directly or through gateways to a blockchain network. The IoT nodes could either be regular sensors with gateway nodes functioning as blockchain nodes (BNs) or they can be integrated with the blockchain itself. As shown in the figure, in a general IoT-blockchain architecture, the data collected from the IoT nodes is forwarded to the blockchain in the form of transactions. These transactions are then validated by a group of miners and added to the blockchain network, making the stored data secure and impenetrable. Some existing methods of integration strategies are summarized in the following subsections based on how IoT nodes are combined together with blockchain.

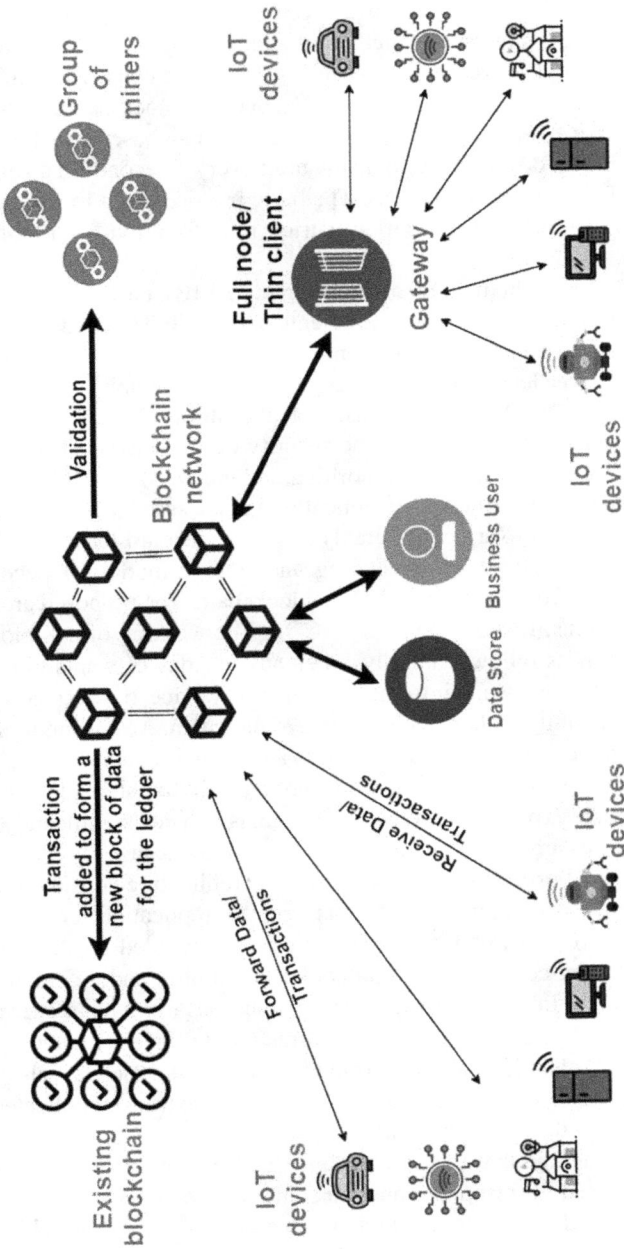

FIGURE 2.2 Different approaches of blockchain integration in IoT.

2.6.1 LIGHTWEIGHT IoT NODES AS THIN CLIENTS

IoT end devices that are always on can function as thin clients in relaying the transactions to full BNs. In this case, IoT gateways can function either as full BNs or thin clients. Danzi et al. [21] have used a method of aggregation at the BNs to reduce the amount of data to be transferred to the IoT nodes. Resource-constrained IoT devices store only a small subset of data from the BNs, such as block headers, when they synchronize with the BNs. IoT devices are connected via a base station to the BNs. The information is sent downlink periodically after every T seconds in an aggregated manner to the IoT devices from the BNs. The basic protocol used for blockchains is the Ethereum protocol, which uses Merkle-Patricia trees for providing proof of inclusion for the modified data.

An efficient lightweight integrated blockchain (ELIB) model is presented by Mohanty et al. [22] for meeting the requirements of the IoT. The model is applied to a resource-constrained smart home environment to verify its credibility. The ELIB model functions at three levels, namely certificateless cryptography (CC), distributed throughput management (DTM), and consensus algorithm. The CC method is used to reduce the overhead associated with the security of new blocks. The creation of new blocks is restricted by consensus algorithm and the DTM method takes control of altering specific system variables dynamically to make sure that the throughput of public blockchain does not vary considerably from the standard load.

Wang et al. [23] proposed a data-sharing and storage model for decentralized access to data derived from the FISCO-BCOS blockchain. The proposed architecture consists of applications, IPFS, and FISCO-BCOS. The working of this blockchain-based IoT network is as follows: initially, when any IoT device wants to join a network, it provides the necessary information, such as device ID, service type, and serial number, for identity authentication by the route chain. According to this identity authentication, the route chain allocates it to a specific set chain. The application then gathers data from the IoT device and encrypts it for storage at the IPFS. Therefore, the storage expenditure of the blockchain is reduced significantly because only IPFS file indexes are stored rather than the whole raw data.

Pohrmen et al. [24] proposed an IoT-blockchain architecture consisting of application, infrastructure, and control layers. User specific application services, such as smart railways, smart grids, and smart campuses, are provided by the application layer. The control layer consists of several connected mining nodes containing high computational powers. In this level, high-security methods like asymmetric cryptography can be used (global blockchain). The infrastructure layer is made up of networks like WSN, IoT, and LTE and networking elements like routers, sensor nodes, and switches. The network element uses local blockchain to transfer collected data to the control layer from the network. When the packet is forwarded, security methods should be implemented so that they are not modified by any intruder. Lightweight hash functions are used for resource-constrained network elements.

A blockchain-based IoT system called BeeKeeper has been proposed by Zhou et al. [25]. In this threshold-based system, three parties function together, namely a leader, its devices, and certain servers. Initially, the leader performs a BeeKeeper protocol which is a (t, n) threshold protocol having complete control over several devices. The servers have the potential to do homomorphic computations over the data encrypted on the blockchain sent by the devices. However, if honest servers are more than n–t, the servers do not have the capability to learn from the data. A leader sends a query through transactions in the blockchain to the servers when it wants a result from the encrypted data. The sensors that are active can respond to the query with the appropriate encrypted data through blockchain transactions. The encrypted responses can be decrypted only by the leader because the decryption key is only available to the leader. The desired outcome can be obtained by the leader if it collects t or more correct responses. Then, the corresponding servers will automatically gain some reward through the smart contract with the leader.

Li et al. [7] proposed a scheme for storage and protection of IoT data using blockchain. Since IoT devices are low-power devices with constrained resources, edge computing is used in this method to carry out computations at the network edge on behalf of IoT devices and store the data in distributed hash tables (DHTs). Before forwarding this data for storage to the DHT, the edge device announces this data by posting a transaction to the blockchain, stating that this data belongs to a certain IoT device. This transaction is then verified by the blockchain and details like the ID of the IoT device and storage address are recorded. Similarly, when data is required by an IoT device from the DHT, it will first post a transaction, which will be authenticated by the blockchain. CC is used as an authentication method for the identification of an IoT device. The drawback of CC is overcome through the use of blockchain's public ledger system, which helps to broadcast the public key of any IoT device in a convenient manner.

NormaChain [26] is a transaction settlement scheme based on blockchains for IoT e-commerce. The regulators need certain information about users' transactions to keep a check on their legitimacy. This dilemma of choosing between privacy and legitimacy is solved by NormaChain through the use of searchable encryption (SE). SE is a method that allows the regulators to search for any particular keywords from the encrypted data without uncovering the whole plaintext. This helps in searching for any criminal manifests while keeping the privacy of the users intact. The authors propose a modified version of the public key encryption with keyword search (PEKS) method, which is decentralized without the need for any central supervision, unlike the original scheme. This supervision is distributed among n parties and the target keyword can be searched only when all the n parties authorize it simultaneously. Therefore, this scheme tolerates corruption up to collusion of n − 1 parties. This scheme is designed to overcome the efficiency issue for e-commerce-based IoT. A layered blockchain is used in which each layer has a different role to achieve scalability and greater transaction speed. Experimentally, this scheme can achieve a supervision accuracy of 100% and the average number of transactions per second is 113.

Xu et al. [27] proposed a non-repudiation service provisioning approach for industrial IoT based on blockchain for evidence recording and providing service. Each service required by the client is divided in advance into two non-executable sections. They are delivered separately through off-chain and on-chain channels after proper recognition of evidence submissions. Specifically, a major portion of the program is conveyed privately through off-chain channels after submitting valid evidence to the blockchain and the remaining part will be posted on the blockchain after recognition of valid evidence of the preceding steps. Service verification methods used in this chapter for industrial IoT do not require complete program codes due to the lightweight cryptographic techniques and homomorphic hash solutions used. Moreover, fair automated smart contracts are used to aid the industrial IoT clients and service providers in solving service disputes efficiently.

A blockchain-based analytical model for wireless IoT systems has been proposed by Sun et al. [28]. The network consists of two types of nodes, namely full function nodes (FNs) and IoT transaction nodes (TNs). TNs are the traditional IoT devices with limited resources and are of two kinds, idle TNs and active TNs, based on whether they are currently transmitting data or not. FNs are the nodes which have sufficient memory and computational powers. They can implement blockchain protocols fully by taking charge of data storage, transaction confirmation, and forming new blocks. FNs are connected through wireless communications with the TNs and via high data links with each other. Information between TNs is transmitted in the form of transactions, which are validated and written in blocks by FNs. Moreover, they have also deduced a probability density function (PDF) of signal-to-interference-plus-noise ratio for transferring data from an IoT to an FN. The security of the proposed model is analysed under different attacks, that is, random FN, random link, and eclipse attacks. A random FN attack leads to the monopolization of some FNs randomly, thus inhibiting the communication of FNs with TNs. A random link attack is caused by unstable links leading to some being blocked or attacked randomly. An eclipse attack leads to the isolation of a TN from the network due to the monopolization of all the victim's uplink and downlink connections by the attacker. The model proved to be valid under these attacks.

BPIIoT [29] is a lightweight blockchain-based platform designed to solve the problems of trust, security, and island connection in the industrial IoT construction process. BPIIoT consists of an on-chain as well as an off-chain network to manage latency and loads in the network. The on-chain network performs all the transactions, such as programmable licenses and digital signatures, while the off-chain network can solve problems of complex processing and storage. Involvement of third parties is avoided by using SMPC (secure multi-party computation) by the on-chain network. Data processing and computations are divided among several nodes that perform the given tasks without revealing the information. The IoT nodes act as lightweight devices consisting of four layers: application, service, network, and device layers. The service layer supports services for blockchain, such as smart contracts and asymmetric encryption, and other general services, such as controller service and I/O interface. The network layer is used for communication between nodes and the blockchain. The device layer consists of actuators and sensors that are deployed in the given field. These IoT devices connect blockchain to the machines through a plug-and-play solution. IoT devices make it possible for the machine to post data to the blockchain, and send and receive transactions in the blockchain.

2.6.2 IoT Gateways as Blockchain Nodes

IoT gateways can function either as a full BN or a thin client. An IoT gateway can route data and verify integrity of data when it functions as a full BN. If used as a thin client, an IoT gateway stores only a few relevant data parts. Mehedi et al. [30]

proposed a reliable and polished blockchain–IoT infrastructure, which represents a change from centralized systems and memory overheads while maintaining privacy and security effectively. They used the standard IoT infrastructure along with decentralized blockchain technology for storing and accessing the data. For integration of blockchain and IoT, they used Ethereum as a blockchain platform and terminal devices enlisting the network technology. Whenever a request is made for storing a transaction, the proposed approach uses the distributed ledger, which gets executed and stored on its own. To protect the users' identities, the terminal devices are organized in a better way. Loukil et al. [31] designed a semantic IoT gateway that improved control over the private resources along with providing protection for collected personal data. This is done by first matching the terms and conditions of service at the customer's end with the privacy preferences at the data owner's end and then inducing a policy adaptable to the described conditions. This privacy policy is converted into smart contracts. Then, in order to host the generated smart contracts, the set of resources are connected to a decentralized network using blockchain technology. There are nine core components of this blockchain framework, namely public IoT network, private IoT network, public blockchain, smart contract, private ledger, transaction, semantic IoT gateway, storage node, and local storage. The semantic IoT gateway binds together the blockchain network, the actuators, and the IoT terminals. When the experiment was performed in the real world, use-case data showed positive results and proved that custom-generated smart contracts can be added to the blockchain technology with a high success rate.

A credit-based mechanism has been proposed by Huang et al. [32] that ensures better efficiency for simultaneous transactions and confirms system security. For ensuring the confidentiality of sensitive data, a mechanism for managing the data authority is defined that controls the sensor data access. The mechanism is designed on structured blockchain based on directed acyclic graphs, which gives better performance and improved throughput compared with chain-structured blockchain like Satoshi-style blockchain. The case study was conducted for a smart factory and the system was implemented on Raspberry Pi 3 Model B. The architecture design of a smart factory consists of four major components, namely tangle network, wireless sensors, managers, and gateways. The designed architecture is impenetrable to various attacks such as DoS, distributed denial-of-service (DDoS), and Sybil. Sybil attacks generally occur in a peer-to-peer network where the attacker creates multiple active identities to hugely influence the network. To guarantee the optimal trade-off between system security and transactional efficiency, a proof-of-work mechanism has been designed, which ensures that the honest nodes always devour a limited number of resources while enforcing the malicious nodes with increased attack cost. To ensure confidentiality, the authors designed a mechanism for data authority management in which the nodes that collect sensitive data are given the secret key by managers, and with the help of this key, sensor data is encrypted before being posted on the blockchain.

Biswas et al. [33] have used a network of local peers which narrows the gap between blockchain peers and IoT devices. Without affecting the transactional validation policy followed by peers at both the local and global level, the number of transactions entering the global blockchain is restricted using a local ledger.

The authors proposed a framework based on blockchain technology for IoT which considers both the inter and intra transactions for the corresponding organization. Each IoT device is registered by the certification authority and associated with one of the organizations. Instead of using peers belonging to a global blockchain, a local peer was structured to achieve the interaction with peers belonging to global blockchain network. The designed framework aims to handle the indirect rise in transactions per second for the global blockchain network and increase in ledger storage requirements at peer level. The size of ledger is limited under this framework and is distributed between local and global peers. The transactions between two organizations are validated via a global blockchain network that provides 100% peer validation. In this work, they clearly demonstrated that if the issue of scalability is not addressed, blockchain and IoT cannot be integrated, and that creating a network of local peers allowed the blockchain ledger to spread across all the peers and hence improved scalability.

A blockchain connected gateway is designed by Cha et al. [34] to maintain the privacy preference of IoT devices securely and adaptively within the blockchain network. It can prevent the leakage of sensitive data by ensuring that the data is not accessed without the user's permission. There are three major participants in the proposed framework, namely the IoT device administrator, gateway administrator, and end user. The IoT device administrator stores the device information along with the device's privacy policy at the blockchain network before the user gains access to it. The list of attributes uploaded on the blockchain network includes the device's name, manufacturer's information, device description, and device images list, and the privacy policy includes information related to preference, policy identifier, and so on. The signature scheme proposed by the authors is robust and has interacting skills similar to DLP based on elliptical curves. Each security component of this signature scheme has been implemented and tested on Raspberry Pi 3 Model B and the computational cost for each security component has been calculated. The results show that while legacy devices are in use, the proposed framework increases the trust among IoT applications and improves user privacy.

Badr et al. [35] designed a novel protocol named pseudonym based on encryption for providing privacy to patients' data available in the e-healthcare system. The encryption mechanism is blockchain-based in which high-end different authority encryption techniques are used for securing the patients' confidential data. The public blockchain tier between the healthcare cloud providers and the blockchain tier handling the sensors on patients' bodies along with the patients' system on the platform are considered in this approach. The work elevates the anonymity factor in patients' data by considering blockchain as the anonymity enhancement technology using the multitier architectural model which prevents the system from various attacks such as block enquiry infringement. The solution was evaluated using the MIRACL library, which keeps track of the processing time for all the functions executing within the communication channel. The architecture contains three tiers, where the first tier shows how all the sensor devices are connected to the patient through a gateway or aggregator. The second tier analyses the distribution of the ledger and handles communication within the health record members and provider. In tier three, compliance with the cloud providers is considered and analysed. This framework can handle some of the security vulnerabilities but not all. Therefore, in the future, this model should be modified to handle larger clusters of security issues.

2.6.3 IoT Nodes Integrated with Blockchain Clients

An IoT battery-powered device may be integrated directly with a blockchain client. This allows blockchain features to be embedded in IoT devices themselves for direct interaction between them. A multilevel blockchain system (MBS) is proposed by Mbarek et al. [36] to secure an IoT that uses mobile agents to enforce the flexibility and speed of transactions in the blockchain. Mobile agents roam throughout the network of IoT devices to aggregate useful data and generate hashed blocks of data, reducing time delays and solving other issues like scalability and synchronization. MBS consists of three hierarchical levels through which IoT devices can send their data securely: micro-level consisting of IoT devices, meso-level consisting of cluster heads, and macro-level consisting of the blockchain platform. The MBS platform is made up of four entities: the IoT device (collects and transmits data), ordering service (accepts transactions and creates blocks), endorsing peers (checks validity of smart contracts), and committing peers (runs validation). It includes meso, macro, and micro agents with different roles and locations in the architecture. Simulation is done using Hyperledger Fabric with 1,000 nodes and the end results are satisfactory in terms of energy consumption and response time.

Qian et al. [37] divide IoT into three parts, namely the network layer, application layer, and perception layer, and propose a security scheme for IoT using blockchain by considering the security issues in these layers. The application layer, consisting of smart homes, smart healthcare, and automatic driving, includes access and authentication control, privacy protection, and software handling. The network layer consists of low-power WANs and mobile networks. The perception layer requires security of devices, authentication, and access control and consists of IoT gateway and terminal devices. To manage the security and other issues of the IoT, blockchain-based platforms for IoT devices can be constructed along with the integration of cloud services. This structure, consisting of union nodes, IoT devices, cloud providers, and so on, communicates through high-speed links. Links between IoT and blockchain devices can be secured through authentication techniques to guarantee reliability. They have also discussed two open issues, namely identity verification and machine learning-based monitoring of abnormal network traffic.

A blockchain-based IoT structure is proposed by Wang et al. [38] using smart contacts, which aids users in keeping complete control over their useful data and also on how it is used by third-party clients. The given system model consists of three entities: aggregates are the users owning IoT devices who post transactions to the blockchain to publish data or grant permissions; subscribers (third parties) want to access the data posted on the blockchain by issuing transactions; and vendors are the IoT devices' manufacturers who are liable for producing official images of firmware. All these three entities are recognized through public–private key pairs when they want to communicate via the blockchain network. Aggregators store their published data in the off-chain network using content-based addressing. A hash is calculated for each data piece corresponding to the address of the data, which is used as an index for data search and retrieval. Two smart contracts, namely firmware update and access control, are introduced for controlling updates and providing access permissions. This blockchain-based update scheme of firmware ensures that the IoT devices are not tampered with and are designed through authentic firmware.

Another hierarchical structure of blockchain for tamper-proof storage and retrieval of data in IoT systems is discussed by Angin et al. [39]. In this architecture, along with the resource-constrained IoT nodes, some additional devices are used for "data collection" which have more storage and computational power. This model ensures that the data is sent securely to the edge servers through the resource-constrained devices for data verification. An authentication and access control method for IoT is proposed by Ourad et al. [40] based on a distinct blockchain-dependent architecture. The authentication process is performed through smart contracts. If found valid, the sender's address and access token are broadcast by the smart contract through which the user can receive this information. A package is then crafted by the user and signed using the Ethereum private key. The authors have shown that this method outperforms existing methods in terms of decentralization and tamper-proof records. This approach can also withstand attacks attempted to guess credentials through brute force and control legitimate sessions.

Bubbles of trust [5] is an authentication method for IoT devices based on public blockchains and smart contracts. In this approach, secure virtual zones called bubbles of trust are created where each device trusts only the devices within its zone. Each zone is inaccessible and protected from non-member devices. Communications in this network are through transactions validated by the blockchain. In the initialization phase, a master device that owns a public–private key pair is designated. All objects within the system are called followers. Each follower is given a ticket that contains an objectID (identifier of follower), groupID (identifier of object's bubble), pubAddr (public address of follower), and a signature. The master of the bubble initiates a transaction containing the identifier of the master and the group created. This transaction is validated by the blockchain to check the uniqueness of identifiers. After the creation of a bubble, the followers send transactions to get linked to their bubble. The follower's identifier is also verified and validated by the blockchain using smart contracts. This approach satisfies the requirements of IoT in terms of its security, cost, and efficiency when implemented using Ethereum and C++ language.

2.6.4 IoT Nodes as Regular Sensors

IoT devices with insufficient resources to tolerate any additional logic function as regular sensors. They simply collect the data and forward it to the blockchain structure through the gateway. Dorri et al. [41] showed that blockchain can be designed as scalable as well as lightweight by optimizing it with respect to the requirements of the IoT with proper end-to-end security. They proposed a DTM algorithm which lowers the delay and processing overhead for mining. The cluster heads deploy the distributed trust technique, such that the verification of new blocks does not result in high processing overheads. To handle scalability, all the overlay nodes, including service providers, cloud storage, and IoT devices, are made a part of clusters and the blockchain is managed by the cluster heads only. To reduce the packet overhead as well as memory footprints, the data of all the nodes is stored off-chain in cloud storage. Each cluster head tries to develop trust with the other cluster heads by validating them on the basis of generated new blocks using the distributed algorithm of trust. By considering the transaction load of the network, the DTM algorithm ensures that throughput of the blockchain remains stable as long as system parameters are handled dynamically. Dorri et al. also performed analysis of the proposed approach under eight significant cyber-attacks and estimated the likelihood of these attacks along with their respective defence mechanisms, which ensured that the algorithm applies to all of them. The simulation was performed on NS3 and the results showed that, in comparison to other approaches, this approach provides higher scalability and reduces the overall delay and packet overhead.

A highly scalable and distributed access management system is proposed by Novo [42] using blockchain technology. The author compares the existing approaches with the proposed approach and illustrates that when it is tested and analysed on different configurations of the blockchain system, the delay was reduced and the throughput was improved. When the sensor nodes are connected through multiple hubs, this approach works better for horizontal scalability and gives better results. The architecture for this model was designed keeping in mind various parameters like concurrency, mobility, lightweight nature, accessibility, and scalability. It comprises various components where each component performs a specific task and all of them when allied together will result in complete working of the system. The sensor network, agent-node, managers, blockchain network, smart contract, and management hub are the various components of this architecture. To test and retrieve the best results, they performed the experiment multiple times using Docker on the Ethereum platform. The results were almost similar in each of the trials and showed positive results.

An access management scheme is proposed by Ding et al. [43] based on attributes for IoT systems. Blockchain is used to store the attribute's distribution to maintain data integrity and avoid single points of failure. Two main entities used in this approach are IoT devices and attribute authorities. The attributes' authorities distribute attributes and manage the blockchain. Attribute-based access extracts the identities (roles) into an attribute set managed by the attribute authorities. These authorities maintain a public ledger jointly using a consensus mechanism. The attributes are authorized and posted to the blockchain through "transactions". For the registration of IoT devices into the network, the attribute authorities create a public–private key pair. The IoT devices collect and share the data in the network. They do not take part in verifying transactions and can only view the blockchain. The access control mechanism is simplified through the use of basic signature and hash techniques to make this approach efficient for resource-constrained IoT systems.

BB-DIS (blockchain and bilinear mapping-based data integrity) is proposed by Wang and Zhang [44] for data integrity of large-scale data of IoT systems. The framework of BB-DIS includes four entities, namely cloud service providers (CSPs), data consumer devices (DCDs), data owner devices (DODs), and smart contracts. Smart contracts used in this approach are of different kinds and they are used to verify the data integrity in the blockchain. DCDs and DODs are used to produce key pairs at the time of initialization of blockchain. CSPs can provide mining services by

TABLE 2.1

Review of blockchain and IoT integration techniques

Title	Year	Method of Integration				Achievements
		1	2	3	4	
Novo [42]	2018				✓	Horizontal scalability
Pohrmen et al. [24]	2018	✓				Enhances security using lightweight cryptographic techniques
Zhou et al. [25]	2018	✓				Fewer computational and memory resources used; fault-tolerant
Hammi et al. [5]	2018			✓		Cost-efficient; resiliency toward attacks; scalability
Li et al. [7]	2018	✓				Certificateless cryptography combined with blockchain; data trading
Cha et al. [34]	2018		✓			Accords users privacy preferences for IoT devices
Ourad et al. [40]	2018			✓		Secure communication; authentication; traceability
Biswas et al. [33]	2018		✓			Scalable local ledger; reduction in the ledger size and block weight
Qian et al. [37]	2018			✓		Low latency and high throughput
Loukil et al. [31]	2018		✓			Flexible and dynamic; privacy preservation
Badr et al. [35]	2018		✓			Privacy preservation for patients; high efficiency and security
Liu et al. [26]	2018	✓				Increased system scalability and transaction efficiency; crime-traceability; 100% supervision accuracy
Bai et al. [29]	2019	✓				Lightweight network; used for smart predictive maintenance and maintenance service sharing
Mehedi et al. [30]	2019		✓			High availability, security, and privacy; less memory overhead
Huang et al. [32]	2019		✓			Transaction efficiency; honest nodes have less power consumption
Wang et al. [38]	2019			✓		Supports arbitrary and different mining rates; improved analytical tractability; ergodic model
Mbarek et al. [36]	2019			✓		Improved flexibility and speed of blockchain; reduced response time and energy consumption
Dorri et al. [41]	2019				✓	Scalability; end-to-end security; reduced overheads and delay; highly fault-tolerant
Mohanty et al. [22]	2020	✓				Reduced processing time and low energy consumption

functioning as miner nodes and receiving the required rewards. Storage services like Microsoft Azure and Amazon S3 are also provided by CSPs. After dividing the IoT data into shards, BB-DIS generates homomorphic verifiable tags (HVTs) for verification of sampling. Edge computing is used to reduce the computation and communication costs by pre-processing the large-scale generated IoT data.

A distributed key management architecture is proposed by Ma et al. [45] to reduce the latency and satisfy the scalability, hierarchical access control, and decentralization requirements of IoT. In this architecture, the blockchain mechanism is governed by the security access managers (SAMs) instead of central authorities. SAMs are used for storing the logical topology, whereas the blockchain is used to store the key management activities. Moreover, multi-blockchains are used by cloud managers to reduce latency and support scalability operations. This multi-blockchain approach improves the scalability and system performance, as shown through simulation results.

Table 2.1 summarizes the approaches of integration where the cases 1, 2, 3, and 4 in the column "Method of Integration" depict the classification method as given in Section 2.6.

2.7 CONCLUSION AND FUTURE WORK

This chapter summarizes the need and usage of blockchain technology to overcome the security risks and scalability issues of IoT devices. The fundamentals of blockchain and IoT are discussed along with the applications of blockchain-based IoT architecture. A detailed study of the state-of-the-art integration approaches of blockchain and IoT is then presented and classified according to the role IoT nodes exhibit in the integration. However, blockchain in IoT has not yet been exploited to its full potential and research in this direction is still in its early phase. Research areas like smart manufacturing, smart energy, and data trading still need a lot of attention and the integration challenges like scalability need to be addressed. Blockchain has the potential to eliminate the use of trusted intermediaries and centralized systems and make the IoT systems a fully secure, distributed, and scalable technology if used efficiently and plenteously.

REFERENCES

[1] F. A. Alaba, M. Othman, I. A. T. Hashem, and F. Alotaibi, "Internet of things security: a survey," *Journal of Network and Computer Applications*, vol. 88, pp. 10–28, 2017.

[2] A. Wang, R. Liang, X. Liu, Y. Zhang, K. Chen, and J. Li, "An inside look at IoT malware," in *International Conference on Industrial IoT Technologies and Applications*, pp. 176–186. Springer, Cham, Switzerland, 2017.

[3] H. Kim and E. A. Lee, "Authentication and authorization for the internet of things," *IT Professional*, vol. 19, no. 5, pp. 27–33, 2017.

[4] C. Kolias, G. Kambourakis, A. Stavrou, and J. Voas, "DDoS in the IoT: Mirai and other botnets," *Computer*, vol. 50, no. 7, pp. 80–84, 2017.

[5] M. T. Hammi, B. Hammi, P. Bellot, and A. Serhrouchni, "Bubbles of trust: a decentralized blockchain-based authentication system for IoT," *Computers & Security*, vol. 78, pp. 126–142, 2018.

[6] O. J. A. Pinno, A. R. A. Gregio, and L. C. De Bona, *"Control chain: blockchain as a central enabler for access control authorizations in the IoT,"* in *IEEE Global Communications Conference*, Singapore, pp. 1–6, 2017.

[7] R. Li, T. Song, B. Mei, H. Li, X. Cheng, and L. Sun, "Blockchain for large-scale internet of things data storage and protection," *IEEE Transactions on Services Computing*, vol. 12, no. 5, pp. 762–771, 2018.

[8] M. A. Khan and K. Salah, "IoT security: review, blockchain solutions, and open challenges," *Future Generation Computer Systems*, vol. 82, pp. 395–411, 2018.

[9] A. Dorri, S.S. Kanhere, R. Jurdak, and P. Gauravaram, "Blockchain for IoT security and privacy: the case study of a smart home," in *2017 IEEE International Conference on Pervasive Computing and Communications Workshops (PerCom Workshops)*, Kona, HI, pp. 618–623, 2017.

[10] K. Biswas and V. Muthukkumarasamy, *"Securing smart cities using blockchain technology,"* in *IEEE 18th International Conference on High Performance Computing and Communications; IEEE 14th International Conference on Smart City; IEEE 2nd International Conference on Data Science and Systems (HPCC/SmartCity/DSS)*, pp. 1392–1393, 2016.

[11] K. N. Griggs, O. Ossipova, C. P. Kohlios, A. N. Baccarini, E. A. Howson, and T. Hayajneh, "Healthcare blockchain system using smart contracts for secure automated remote patient monitoring," *Journal of Medical Systems*, vol. 42, no. 7, p. 130, 2018.

[12] P. Zhang and M. N. K. Boulos, "Blockchain solutions for healthcare," in *Precision Medicine for Investigators, Practitioners and Providers*, pp. 519–524. Elsevier, 2020.

[13] V. Sharma, "An energy-efficient transaction model for the blockchain-enabled internet of vehicles (IoV)," *IEEE Communications Letters*, vol. 23, no. 2, pp. 246–249, 2018.

[14] T. Jiang, H. Fang, and H. Wang, "Blockchain-based internet of vehicles: distributed network architecture and performance analysis," *IEEE Internet of Things Journal*, vol. 6, no. 3, pp. 4640–4649, 2018.

[15] S. A. Abeyratne and R. P. Monfared, "Blockchain ready manufacturing supply chain using distributed ledger," *International Journal of Research in Engineering and Technology*, vol. 5, no. 9, pp. 1–10, 2016.

[16] A. Bahga and V. K. Madisetti, "Blockchain platform for industrial internet of things," *Journal of Software Engineering and Applications*, vol. 9, no. 10, p. 533, 2016.

[17] S. Saberi, M. Kouhizadeh, J. Sarkis, and L. Shen, "Blockchain technology and its relationships to sustainable supply chain management," *International Journal of Production Research*, vol. 57, no. 7, pp. 2117–2135, 2019.

[18] N. Kshetri, "1 blockchain's roles in meeting key supply chain management objectives," *International Journal of Information Management*, vol. 39, pp. 80–89, 2018.

[19] M. Mylrea and S. N. G. Gourisetti, *"Blockchain for smart grid resilience: exchanging distributed energy at speed, scale and security,"* in *IEEE Resilience Week (RWS)*, Wilmington, DE, pp. 18–23, 2017.

[20] N. Z. Aitzhan and D. Svetinovic, "Security and privacy in decentralized energy trading through multi-signatures, blockchain and anonymous messaging streams," *IEEE Transactions on Dependable and Secure Computing*, vol. 15, no. 5, pp. 840–852, 2016.

[21] P. Danzi, A. E. Kalør, C. Stefanovi'c, and P. Popovski, "Delay and communication tradeoffs for blockchain systems with lightweight IoT clients," *IEEE Internet of Things Journal*, vol. 6, no. 2, pp. 2354–2365, 2019.

[22] S. N. Mohanty, K. Ramya, S. S. Rani, D. Gupta, K. Shankar, S. Lakshmanaprabu, and A. Khanna, "An efficient lightweight integrated blockchain (ELIB) model for IoT security and privacy," *Future Generation Computer Systems*, vol. 102, pp. 1027–1037, 2020.

[23] Y. Wang, C. Wang, X. Luo, K. Zhang, and H. Li, "A blockchain-based IoT data management system for secure and scalable data sharing," in *International Conference on Network and System Security*, pp. 167–184. Springer, Cham, Switzerland, 2019.

[24] F. H. Pohrmen, R. K. Das, W. Khongbuh, and G. Saha, "Blockchain-based security aspects in internet of things network," in *International Conference on Advanced Informatics for Computing Research*, pp. 346–357. Springer, Cham, Switzerland, 2018.

[25] L. Zhou, L. Wang, Y. Sun, and P. Lv, "Beekeeper: a blockchain-based IoT system with secure storage and homomorphic computation," *IEEE Access*, vol. 6, pp. 43472–43488, 2018.

[26] C. Liu, Y. Xiao, V. Javangula, Q. Hu, S. Wang, and X. Cheng, "Normachain: a block chain-based normalized autonomous transaction settlement system for IoT-based e-commerce," *IEEE Internet of Things Journal*, vol. 6, no. 3, pp. 4680–4693, 2018.

[27] Y. Xu, J. Ren, G. Wang, C. Zhang, J. Yang, and Y. Zhang, "A blockchain-based non-repudiation network computing service scheme for industrial IoT," *IEEE Transactions on Industrial Informatics*, vol. 15, no. 6, pp. 3632–3641, 2019.

[28] Y. Sun, L. Zhang, G. Feng, B. Yang, B. Cao, and M. A. Imran, "Blockchain-enabled wireless internet of things: performance analysis and optimal communication node deployment," *IEEE Internet of Things Journal*, vol. 6, no. 3, pp. 5791–5802, 2019.

[29] L. Bai, M. Hu, M. Liu, and J. Wang, "Bpiiot: a light-weighted blockchain-based platform for industrial IoT," *IEEE Access*, vol. 7, pp. 58381–58393, 2019.

[30] S. T. Mehedi, A. A. M. Shamim, and M. B. A. Miah, "Blockchain-based security management of IoT infrastructure with ethereum transactions," *Iran Journal of Computer Science*, vol. 2, no. 3, pp. 189–195, 2019.

[31] F. Loukil, C. Ghedira-Guegan, K. Boukadi, and A. N. Benharkat, "Semantic IoT gateway: towards automated generation of privacy-preserving smart contracts in the internet of things," in *OTM Confederated International Conferences "On the Move to Meaningful Internet Systems"*, pp. 207–225. Springer, Cham, Switzerland, 2018.

[32] J. Huang, L. Kong, G. Chen, M.-Y. Wu, X. Liu, and P. Zeng, "Towards secure industrial IoT: blockchain system with credit-based consensus mechanism," *IEEE Transactions on Industrial Informatics*, vol. 15, no. 6, pp. 3680–3689, 2019.

[33] S. Biswas, K. Sharif, F. Li, B. Nour, and Y. Wang, "A scalable blockchain frame-work for secure transactions in IoT," *IEEE Internet of Things Journal*, vol. 6, no. 3, pp. 4650–4659, 2019.

[34] S. C. Cha, J. F. Chen, C. Su, and K. H. Yeh, "A blockchain connected gateway for BLE-based devices in the internet of things," *IEEE Access*, vol. 6, pp. 24639–24649, 2018.

[35] S. Badr, I. Gomaa, and E. AbdElrahman, "Multi-tier blockchain framework for IoT-EHRs systems," *Procedia Computer Science*, vol. 141, pp. 159–166, 2018.

[36] B. Mbarek, N. Jabeur, T. Pitner et al., "MBS: multilevel blockchain system for IoT," *Personal and Ubiquitous Computing*, pp. 1–8, 2019.

[37] Y. Qian, Y. Jiang, J. Chen, Y. Zhang, J. Song, M. Zhou, and M. Pustisek, "Towards decentralized IoT security enhancement: a blockchain approach," *Computers & Electrical Engineering*, vol. 72, pp. 266–273, 2018.

[38] X. Wang, G. Yu, X. Zha, W. Ni, R. P. Liu, Y. J. Guo, K. Zheng, and X. Niu, "Capacity of blockchain based internet-of-things: testbed and analysis," *Internet of Things*, vol. 8, p. 100109, 2019.

[39] P. Angin, M. B. Mert, O. Mete, A. Ramazanli, K. Sarica, and B. Gungoren, "A block-chain-based decentralized security architecture for IoT," in *International Conference on Internet of Things*, pp. 3–18. Springer, Cham, Switzerland, 2018.

[40] A. Z. Ourad, B. Belgacem, and K. Salah, "Using blockchain for IoT access control and authentication management," in *International Conference on Internet of Things*, pp. 150–164. Springer, Cham, Switzerland, 2018.

[41] A. Dorri, S. S. Kanhere, R. Jurdak, and P. Gauravaram, "LSB: a lightweight scalable blockchain for IoT security and anonymity," *Journal of Parallel and Distributed Computing*, vol. 134, pp. 180–197, 2019.

[42] O. Novo, "Scalable access management in IoT using blockchain: a performance evaluation," *IEEE Internet of Things Journal*, vol. 6, no. 3, pp. 4694–4701, 2018.

[43] S. Ding, J. Cao, C. Li, K. Fan, and H. Li, "A novel attribute-based access control scheme using blockchain for IoT," *IEEE Access*, vol. 7, pp. 38431–38441, 2019.

[44] H. Wang and J. Zhang, "Blockchain based data integrity verification for large-scale IoT data," *IEEE Access*, vol. 7, pp. 164996–165006, 2019.

[45] M. Ma, G. Shi, and F. Li, "Privacy-oriented blockchain-based distributed key management architecture for hierarchical access control in the IoT scenario," *IEEE Access*, vol. 7, pp. 34045–34059, 2019.

3 IoT and Big Data Using Intelligence

Kayal Padmanandam
BVRITH, India

3.1 IoT IN A NUTSHELL

The Internet of Things (IoT) is a system of interconnected devices, machines, applications, and people with a unique device identifier. These systems will transmit data over the network with no human intervention simply through computer interaction. Each component of the IoT is responsible for an action, which altogether makes the job smarter. The device can be an electrical or electronic thing. It receives data in the form of a sensor, signal, or input from an environment. This data is collected and converted to a structured form, stored in the cloud, and used for all types of system-oriented analytics. The human is connected at the other end by means of a user interface system, for instance, a mobile application. Figure 3.1 depicts the vital IoT components.

The IoT device may be a wearable or an embedded device. They have wide applications in consumer, commercial, education, automotive, industrial, and healthcare markets. **Wearable technology** is devices which can be worn on a human body and often include tracking options when connected to the Internet. Wearable technology has the potential to improve lives. It helps humankind in different ways such as regular health monitoring, tracking fitness level, entertainment, planning and scheduling daily routines, and many more that appear trendy and smarter.

Embedded technology is a device with a highly special-purpose computing system. It can be an electrical/electronic/mechanical device with the combination of processor, memory, software, and I/O peripherals. Traditional embedded systems have been designed without the capability of connecting to the Internet, but nowadays, due to increased intelligence and increasing availability of the Internet, they are built with the capability to connect to the Internet through Wi-Fi, Bluetooth, Ethernet, or through any other communication medium. The architecture of the embedded system is depicted in Figure 3.2.

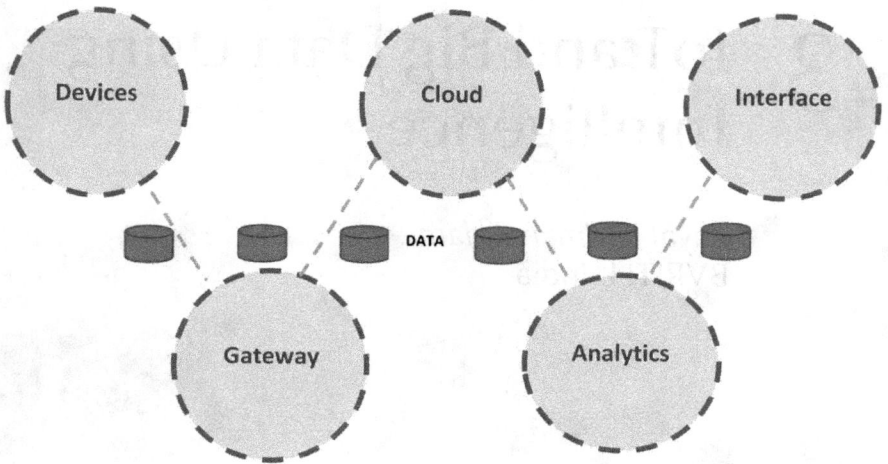

FIGURE 3.1 IoT components.

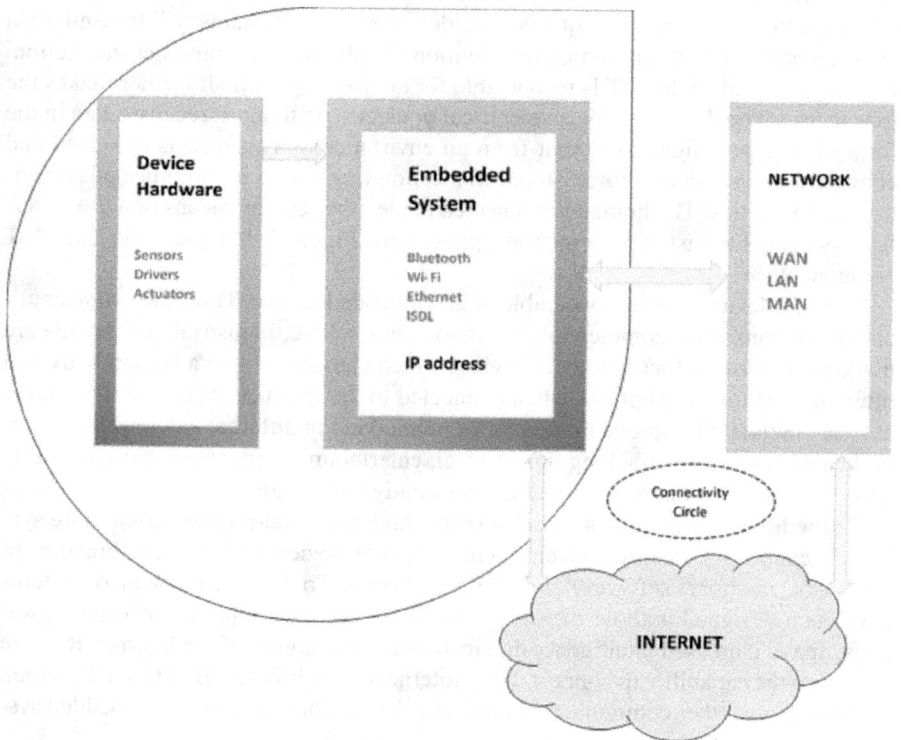

FIGURE 3.2 Architecture of embedded systems.

When these things are on the Internet, we call it an IoT. It is the way we communicate with our technology. However, people's mood of melancholy towards the business of typing, swiping, or managing apps is increasing. Technologies are always disruptive, and the newest way to interact with your system is through voice communication, which is nowadays a bigger fraction of the technology interface included in almost every gadget varying from smartphones, virtual assistants, devices connected to home, automobile solutions, and smart health devices. Google Home mini, Apple's Siri, and Amazon's Alexa are a few of the demonstrated hands-free voice assistants on the market today. Some of the day-to-day tasks of a voice assistant could be adding events to your calendar or items to your to-do list, controlling your smart home devices, making and receiving calls, finding nearby hotels, restaurants, and maps for travel, and checking travel status for your journey.

Here is an example of one the youngest Generation-Z (Gen Z) having a conversation with a voice assistant (VA):

GEN Z: VA, how to draw mummy?
VA: A mummy is a deceased human or an animal whose skin and organs have been preserved.
(URL provided to draw)

GEN Z: 😟 .... I want to draw my mummy.
VA: I am afraid I don't understand.
GEN Z: Even I am afraid of my mummy.

In this comic snip, there are so many things to observe. First the impact of technological influence on an innocent child, the practicality of usage of VA by a child, the belief of the child for learning through IoT, and finally the importance of uttering a word in context.

While this is marked progress, we should not forget that ***"Technologies are always disruptive"*** and so we keep moving on with the next innovation.

3.2   THE BUZZWORD: BIG DATA IN A NUTSHELL

What is big data? Why is it so imperative? Why is the world behind it? We know the data is growing exponentially every nanosecond. The challenge is to turn this data into knowledge. How do you do that? What is the technology needed to handle such exponentially growing data? Where are you storing it? How can you retrieve information? So many questions flash through your mind. There is only one answer to these questions: **big data**.

Big data refers to a tremendously huge data set that can be analysed to unveil the patterns, trends, and associations that can aid better decision-making. Put simply, big data is the key to information and insight about the unimaginably large data set [2]. This insight can be used for predicting fraudulent behaviour, diagnosing a patient's health, categorizing customers' needs, and many more.

Now we admit that big data is big, but it is now time to understand how big. What are the measures that differentiate between data and big data? The data is huge and so processing it using conservative data processing software or techniques is ineffectual. Hence, it becomes mandatory to understand the importance of the big data Three Vs – volume, velocity, and variety.

(a) **Volume – the Size**

Volume refers to the size of the data. Dealing with petabytes (1 million gigabytes), exabytes (1 billion gigabytes), and more needs organized storage management, and software for data analytics. When you think about such requirements, you are definitely working with big data.

(b) **Velocity – the Speed**

Moving from the size of the data, now think about the speed of the data. How fast are the data generated? In terms of a social networking post, for instance, think about the speed of data generated via Twitter messages or a Facebook post. Such platforms require unique processing techniques to handle terabytes, petabytes, and exabytes, and to derive quick insights in real-time.

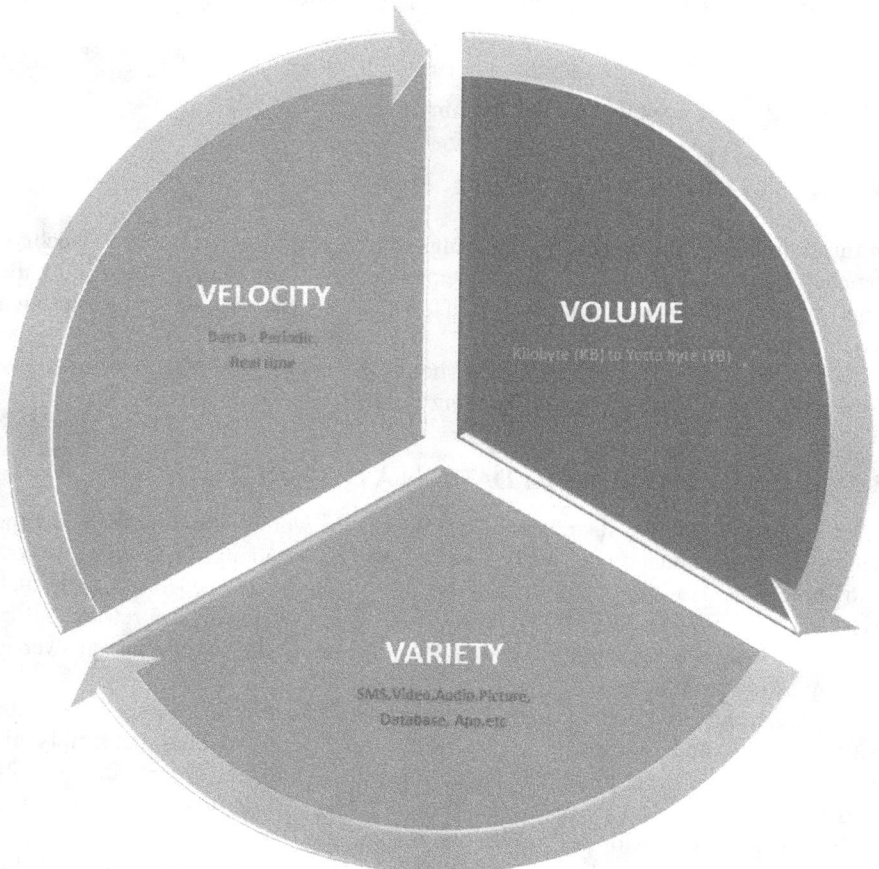

FIGURE 3.3   Basic big data Three Vs.

(c) **Variety – the Format**
When you are already facing the challenges of volume and velocity, variety of data makes it even more complex. Data comes with multi-structured forms – structured data like numeric and text types, semi-structured, and unstructured data like audio, video, images, and mails, for instance. Dealing with such complex data would be tedious without the help of a technological solution software. Figure 3.3 explains about the basic Three Vs of big data.

When an application is domain-specific, it is true that these Three Vs need further contemplation to make a perfect big data model. Therefore, we also have to look for application-specific determinants besides these Three Vs. This is explained by M-Brain [3] in the big data with Eight Vs. Let us have a closer look at the insight provided by the Eight Vs given in Figure 3.4.

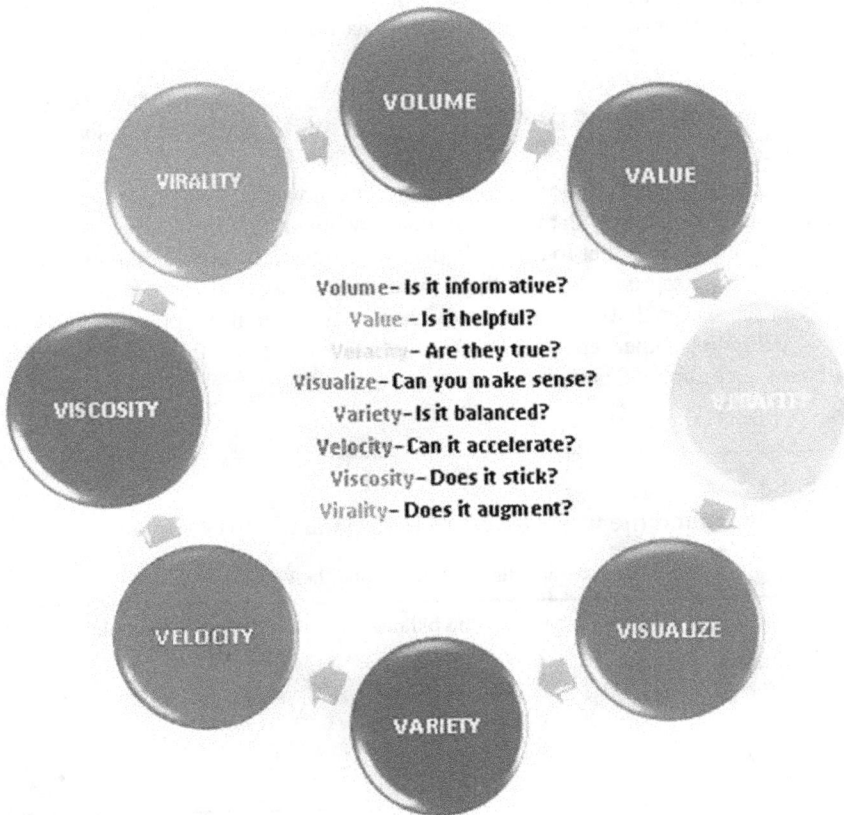

Volume– Is it informative?
Value – Is it helpful?
Veracity – Are they true?
Visualize– Can you make sense?
Variety– Is it balanced?
Velocity– Can it accelerate?
Viscosity– Does it stick?
Virality– Does it augment?

FIGURE 3.4   M-Brain big data Eight Vs.

3.3 IoT VS BIG DATA

It is now clear about the basic principles of big data and IoT. Now it is time to understand the variances and amalgamation of the two giant but different areas of the techno-sphere. The world generates data on override. There is no sign for strike. *Data Never Sleeps*, an article by DOMO, a cloud-based industry partner, illustrates precisely the data generation happening every single minute around the globe in the form of swipes, likes, tweets, and shares. The digitized globe is exploding [3]. Table 3.1 clearly explains the exponentially growing digital data around the world.

The merging of big data and IoT happens at the fundamental compromise with data. Big data collects data from different online and social media sources; the IoT accumulates data from every device around us. Though the underlying characteristics differ, the source is same. Hence, we have to understand the difference between the two, so that our emphasis is on collectively applying the capabilities for a domain-specific, achievable lead.

> "One of the myths about the Internet of Things is that companies have all the data they need, but their real challenge is making sense of it. In reality, the cost of collecting some kinds of data remains too high, the quality of the data isn't always good enough, and it remains difficult to integrate multiple data sources."
>
> Chris Murphy
> Editor, Information Week

Business solutions and industry requirements are nowadays highly reliant on big data. Expenditure on big data management by giant companies has drastically improved. They believe that IoT and big data will impact the global market by delivering value to enterprises through the decreased expense and innovative new avenues. And the fact is that they are already. It is also true that the propagation of big data is pervading other applications like education, research and development, healthcare, e-commerce, and so on.

TABLE 3.1
Digital data around the world – according to January 2019 survey

The Essential Digital Data Around the World

	In billions	Urbanization in %
Total population	7.676	56
User Type	**In billions**	**Penetration in %**
Mobile users	5.112	67
Internet users	4.388	57
Active social media users	3.484	45
Mobile social media users	3.256	42

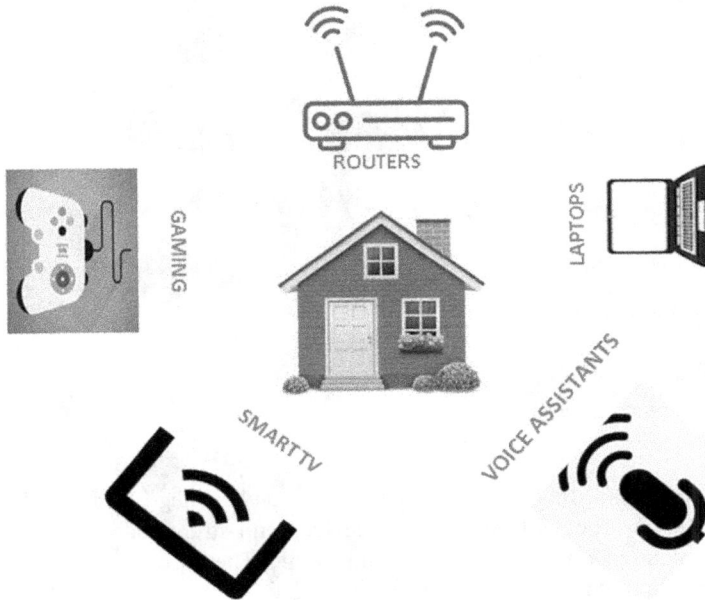

FIGURE 3.5 The IoT and you.

Imagine a world as shown in Figure 3.5, where you and your home are plugged into the Internet. Your home appliances abide your commands, switch on devices automatically when you require them, play music according to your mood, turn it off when you leave, maintain the temperature of your coffee, remind you about your appointments, talk about the news today, alert you with weather and traffic reports, and ultimately drive your car without your interference. This is the world of IoT.

3.4 DATA GENERATION – MACHINE VS HUMAN

The IoT is anticipated to produce 79.4 zettabytes of data with 41.6 billion IoT devices in 2025, according to the International Data Corporation (IDC), which is the premium global provider of market intelligence. You may be looking aghast, but reality has to be accepted! Now the question is where does this data come from? How does it get generated? Is that machine or human?

3.4.1 MACHINE-GENERATED DATA

Data collected by sensors is fundamentally machine-generated data. Sensors are the managers of the physical computational devices in our IoT ecosystem. They are responsible for the act of sensing and acquiring data from the environment. It can be a machine-to-machine interaction or from an IoT device. This data can be structured,

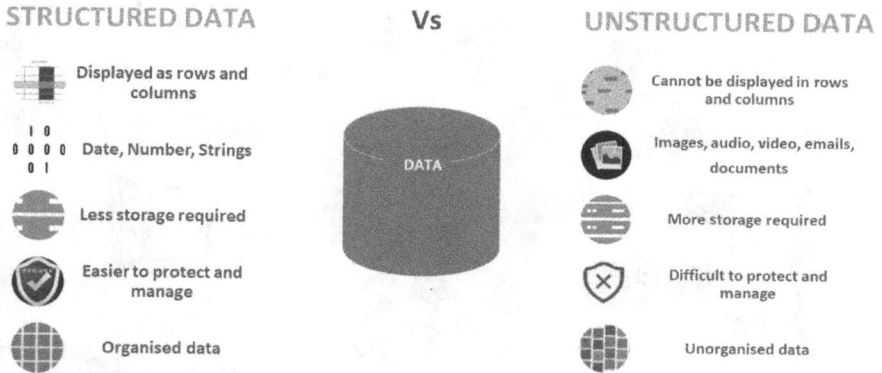

STRUCTURED DATA   Vs   **UNSTRUCTURED DATA**

Displayed as rows and columns

Date, Number, Strings

Less storage required

Easier to protect and manage

Organised data

Cannot be displayed in rows and columns

Images, audio, video, emails, documents

More storage required

Difficult to protect and manage

Unorganised data

FIGURE 3.6   Structured vs unstructured data.

semi-structured, or unstructured. The type of data can be categorized based on data origin. It would be structured if it is from a device, semi-structured if it is from a log file, and unstructured when it is a video or image file. The difference between structured and unstructured data is shown in Figure 3.6. Some examples of machine-generated data can be RFID, GPS output, temperature and other environmental sensing, terrestrial and satellite computer logs, network logs, call records, telemetry which is collected by the government for intelligence purposes, and so on. To define machine-generated data, it is the result of a computer-generated process where human intervention is not involved at all.

HOW?

Big Data and IoT differs
Yet Biddable Across

May be DATA remains
Fundamental for both.

3.4.2   HUMAN-GENERATED DATA

Human-generated data comprises all files and data that we create every day. Again, this data can be structured, semi-structured, or unstructured. For instance, human-generated structured data can be any piece of input data entered into a device. It can be personal details, survey responses, and so on. Human-generated unstructured data can include our emails, social media usage, messages we send and receive, word processing documents, presentation documents, spreadsheets, audio files,

TABLE 3.2
Comparison – Human vs Machine-generated data

Types		Human-Generated	Machine-Generated
Structured	**Internal**	Review scores Talent evaluation Documents	Metrics from log Sales, purchase tracking Measures of progression control
	External	Number of reply tweets Facebook likes Product ratings	GPS for tweets Time log of updates/posts/tweets Smart machinery
Unstructured	**Internal**	Emails, letters, messages YouTube videos Images, voicemails, audio transcripts	Surveillance videos RFID Product scanner
	External	Facebook comments, tweets Product reviews Pinterest images	Satellite videos Scientific data Radar data

video files, and so on. The complexity with human-generated data is the data's metadata. Metadata is defined as the data about the data. Metadata gives the complete picture of the data stored by a human – the type, size, location, access privileges, and so on. Metadata can be much larger than the original data. Researchers can mine actionable intelligence from human-generated data – for example, detecting patterns and trends in the behaviour of the customer, patient, business, and so on – and make decisions and recommendations for long-term solutions. A comparison of human vs machine-generated data is given in Table 3.2 based on the structured and unstructured data formats.

IoT extensively relies on machine-generated data from sensors that are attached to the devices to collect and analyse real-time data for an optimized business solution. They are capable of generating data at higher velocity. Human-generated data also paves the way to augmenting exceptional performance by the IoT system. Many IoT systems rely on both human and machine-generated data depending on the type of environment. For instance, application and business-oriented software is likely to work on human-generated data, whereas societal and environmental-oriented software works with machine-generated data.

3.5 DATA STREAM, MANAGEMENT, AND PROGRESSION USING IoT AND BIG DATA APPROACH

There were times when database analysts worked on a batch processing system. The batch processing system works on known, finite data chunks with multiple Central Processing Units (CPUs) and usually takes a long time for data evaluation. This could not compete nor accommodate the data multiplication that is happening today.

This is an era of big data. Besides volume, the velocity of data is extraordinary to data analysts because the high-speed continuous data has to be collected, processed, and analysed around the clock. A conventional database management system that is monotonous and static in nature could not handle such tremendously complex data in terms of volume, velocity, and veracity. Data streaming, progression, and novel approaches in data handling algorithms should be the primary focus of any data analyst today. The debate on how IoT and big data handle such challenges is always fascinating. The first and foremost difference between IoT and big data is data streaming. Big data collects voluminous data and processes it for insight. There may be a time lag between the data collection and the appropriate insight. On the other hand, in IoT, the collection and processing of data and decision-making happen simultaneously. Consequently, in IoT, there is lot about real-time processing, optimization needs, and regular data management for data-driven decision-making, but both go hand in hand for impactful data analytics. Table 3.3 depicts the differences between batch and streaming processing.

3.5.1 Data Streaming in IoT

We all are very well aware of the database management system (DBMS), which works on three ideologies: data acquisition, data processing, and data storage and management. However, when we work with real-time data, streaming processing, which is the key for data analytics, is enabled by a data streaming management system (DSMS). The differences between DBMS and DSMS are given in Table 3.4. Various real-time devices are capable of producing streaming data. For any real-time data streaming process, Rakusen has recommended the use of techniques like

TABLE 3.3
Batch vs streaming processing

Dimension	Batch Processing	Streaming Processing
Input	Data chunks	Stream of data
Data scope	Finite/known	Infinite/unknown
Computer hardware	Multiple CPUs	Single CPU
Data storage	Stored in memory	No storage
Process time	Several passes	Limited passes
Time	Longer time	Shorter time

TABLE 3.4
DBMS vs DSMS

Database Management System (DBMS)	Data Stream Management System (DSMS)
Permanent data	Volatile data
Random access	Sequential access
One-time query	Continuous query
Minimal update rate	Maximal update rate
No time requirement	Real-time requirement
Query processing	Varied data acquisition
Enquires exact data	Works on inaccurate data too

assessment, aggregation, correlation, and temporal analysis [4]. Because big data is capable of collecting enormous amounts of data, there is a chance of irrelevant data. **Assessment** of data could help exclude such uninteresting data sets from further processing and help save bandwidth by simply discarding them.

Aggregation of data is the process of collecting data and summarizing it. Data collection happens from various sources and the analysis is dependent on the quality and amount of data collected. For instance, aggregation helps businesses by finding out the behaviour of the customer demographic. It can say which products are successful, and how to plan a budget for the business based on each product's selling history. It is also very useful to find any type of fraudulent behaviour happening in the system. We can set alerts for such scenarios, and make business risk-free. Using aggregation in IoT detects events of interest in the real-time data obtained from the device and triggers necessary actions to take place at the right moment. It can also be used to monitor the sensor data from the IoT device to analyse any interesting or risky pattern, or trends that identify a problem or solution to an existing problem. This end-to-end framework, as shown in Figure 3.7, is helps real-time decision-making systems from collection of data to transformation of knowledge.

Correlation of data is the process of identifying attributes that are associated positively or negatively. Multiple streams of data are connected and correlation has to identify the association between these data. For instance, the occurrence of event A is followed by event B. Such prediction is helpful in business, finance, medical diagnosis, and many more situations. Using correlation in IoT enables us to sense the real-time data and its influence on business. It can help a customer to understand that the cab rates increase during peak hours, or it can trigger an alert about machinery based on temperature and take any immediate action required to avoid any kind of damage.

Temporal analysis examines and models the behaviour based on time. It is mostly used in fraud and crime detection as it can predict the pattern of outlier activities with time as the primary element. For instance, the record of the stock exchange can be monitored based on time, and the geologist records the oceanographic pattern for

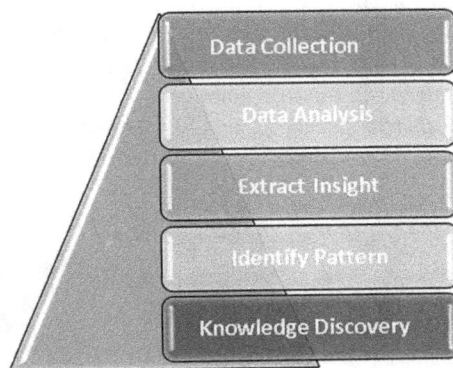

FIGURE 3.7   Data to knowledge framework.

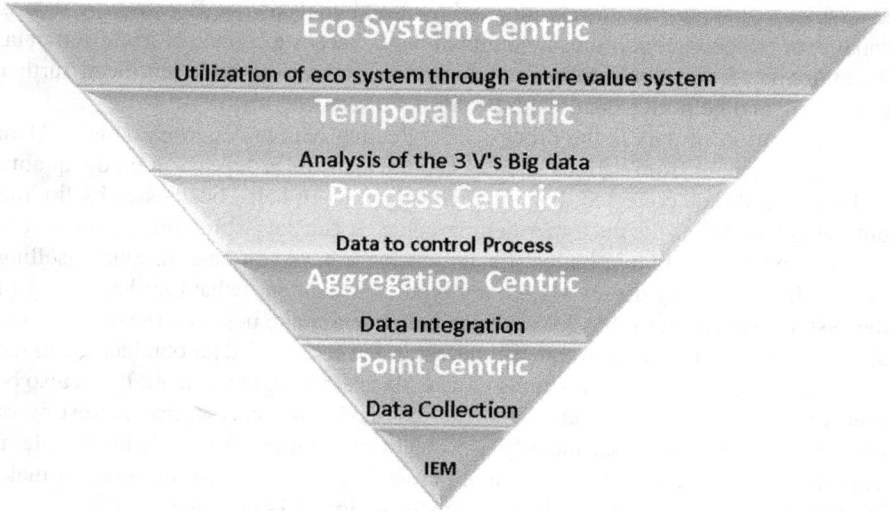

FIGURE 3.8 IoT ecosystem maturity.

coastal threats based on time, which helps to give alerts to coastal people on any emergency evacuation. Ensuring analytical power in IoT increases its belief among people. Any type of travel updates, such as arrival times for metro trains or buses at the station, or school buses reaching the students' stops, can be updated online and help to keep the IoT user comfortable.

By looking into the above abilities, we can understand the supremacy of the data streaming process compared to other conventional data management systems. Data streaming helps us to understand what happened in the past and what is happening now, and it can also be proactive in predicting what will happen in the future, by using predictive algorithms in the streaming process. Using the streaming capabilities, decisions and actions can be taken immediately, which can augment the business situation for a better outcome. The IoT system reaches its maturity after the intelligent use of data analytics. The IoT can translate sensor data, phase by phase, into information, knowledge, and wisdom and finally makes the IoT ecosystem intelligent. A matured IoT ecosystem is depicted in Figure 3.8.

Streams of data from the IoT create a new ecosphere by offering sophistication, ease, connection, intelligence, decisions, and service with enhanced capabilities and ensured optimal insight.

"Like air and drinking water, being digital will be noticed only by its absence, not its presence."
- Nicholas Negroponte[11]

IoT can translate sensor data phase by phase into

3.6   IoT AND BIG DATA WORKING TOGETHER USING INTELLIGENCE

The role of big data begins when IoT collects data at high velocity. To process, store, and understand the real-time data, IoT joins with big data that uses different technologies to bring insight for the data collected. IoT big data has evolved in five stages [5] as shown in Figure 3.9.

Stage 1: Device Connectivity and Data Forwarding

The first and foremost source of data collection is through the device with sensors that are connected via the Internet. The device must be capable of collecting all telematic information, and machine-to-machine (M2M) and sensor data, and store it in a location for further action. The device may have challenges posed by bandwidth limitations but does not stop collecting and storing data. Device and device data collection and storage of collected data are the preliminary steps of IoT which form the foundation for an IoT solution.

Stage 2: Real-Time Monitoring

Necessity is the mother of invention. Only when data is enormous is data processing required. Real-time data monitoring is an important but challenging stage of IoT maturity. As data is collected, it has to be monitored and visualized to suggest outcomes for a problem that can be anything – for instance, suggesting a nearby cafe, alerting about a traffic route, or warning about a vehicle exceeding the speed limit – and adopt a business process for outcomes. Now comes the question of monitoring. Who is monitoring the data? Usually, the real-time system works with a dashboard that shows all basic information, data visualization, and alerting techniques. For instance, the researcher could identify the likelihood of an earthquake occurring before a stipulated time using this real-time dashboard monitoring so that action is taken. Hence, data at this stage is called actionable data. However, the challenge is how humans can monitor such enormous data. The scalability between real-time data and human-dashboard monitoring is inconceivable. The efficacy and accuracy of predictions will be compromised. The solution could be adopting a software-dashboard monitoring system. Moreover, the basic dashboard system might not be effective for complex analysis, so a fully-fledged complex event processing of big data analytics is required for successful real-time monitoring.

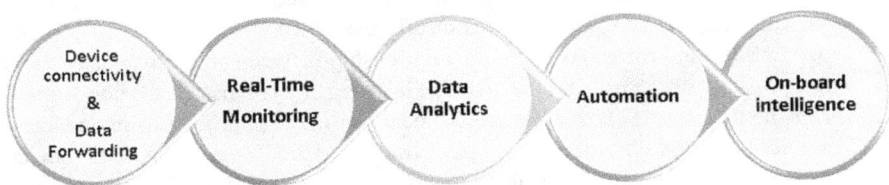

FIGURE 3.9   Five stages in IoT big data.

STAGE 3: AUTOMATED ANALYTICS – BIG DATA ANALYTICS AND COMPLEX EVENT PROCESSING

This stage is the heart of IoT and big data analytics. At this stage, data is transformed into insight, which is followed by prediction and accuracy optimization. As discussed earlier, data can be structured or semi-structured. Hence, data alignment is required to work with different types of data for uniformity in data processing. Numerous techniques and logics are required for an effective data analytics system. First and foremost is the right data for the right prediction and outcome. Only with the right data can we apply complex event processing techniques that can lead to actionable insights. Complex event processing is the place where we apply intelligence to the analytics. It can be a machine learning algorithm based on a clustering technique, regression analysis, or reinforcement learning.

Machine Learning Algorithm

Machine learning is a technique of describing data, acquiring knowledge from that data, and then applying what has been learned to make an informed decision. It provides the system with the ability to automatically learn from experience. When the acquisition of data is perfect, the algorithms can be applied for identifying the trends and patterns in the pool of data collected. This pattern analysis can be helpful in various domains like healthcare, business, education, and many more. For instance, a wearable IoT-based blood pressure monitor can collect the cardiac data of the patient on a chronological basis. This data is analysed by the machine learning algorithm to monitor and identify patterns of any criticality in the cardiac status of the patient.

(a) **Clustering Technique**

Clustering is a reliable communication methodology for devices connected to the network. It requires less energy consumption compared to traditional Mesh terminology. It helps in collaboration as well as for prolonging overall network lifetime. Using an unsupervised clustering mechanism, devices that behave similarly can be identified and grouped for any business decision [6]. It also deals with how the sensors of interconnected devices interact with each other. This interaction is necessary for insightful decision-making. Hence, spending time on the selection of the clustering technique enhances the intelligence of the IoT system.

(b) **Regression Analysis**

Regression analysis is a set of statistical processes for analysing the relationships between a dependent variable (outcome variable) and one or more independent variables (predictor variables). It is mostly used for a distinct process, to make **predictions** and **forecasts** using the relationship between attributes where there is a substantial overlap with the machine learning algorithms. This is how regression helps many IoT devices in the decision-making process. For instance, regression-based control systems have been modelled recently. An IoT device quality of evaluation is determined based on regression [6].

(c) **Reinforcement Learning**

The intelligence of a system is acquired by repetitive learning. The supervised and unsupervised categories of machine learning algorithms have played a vital role in this. Providing huge amounts of training data for the algorithm to learn is not always feasible. On this subject, a new thought-provoking algorithm called reinforcement learning has been taken up by the world today. It is a highly applicable method for IoT scenarios. IoT devices are capable of collecting an enormous amount of data and reinforcement learning can auto convert a quantity of this data into a training data set by using intelligence. Then it starts learning the algorithm using award/penalty combinations whereby correct learning is rewarded and incorrect learning is penalized. This step is repeated until the system learns the maximum possible rewards.

STAGE 4: AUTOMATION

Once the analytics and algorithm are applied, the next foremost complex action is the integration of the same algorithm into the device. For instance, the algorithm can be for a health monitoring system, inventory control, or any support system. The device should be capable of executing end-to-end operations without human intervention in the complex process. From data collection to action performed, the device has to be an intelligent performer. For instance, while collecting data, the machine should understand the amount of data it requires to understand the scenario, where sometimes it requires enormous data to learn whereas at other times it requires less. It is also possible that when a certain condition is met, the system should shut down automatically to prevent any hazardous reaction against the system. Data analytics are a one-off process and are not repeated often. So, the durability of the algorithm is measured after integration. Initially, the integration works smoothly, but as days pass so many internal and external factors may affect the system's effectiveness. We may happen to identify that the rules and the algorithm behave indifferently from the original analysis due to small changes or updates in the software or hardware environment. Automation should be agile to help in such expected and unexpected situations.

STAGE 5: ON-BOARD INTELLIGENCE

On-board intelligence gives the IoT device a complete maturity. It is the time when imaginations and ideas become reality. "Rather than moving the data to the logic, on-board intelligence brings the logic to the data" [4]. When a device acquires such maturity, it is capable of withstanding any predictive failure or device optimization. With on-board intelligence, the device can also perform appreciated functionality when a device connection is lost. The insight gained from analytics should help improve the business process and product development, which is the ultimate true value of IoT.

3.7 WORKING CHALLENGES

The Internet has overhauled the conduct of business, global communication, personal, and occupational interactions. After the rise of IoT, various automated gadgets and applications have appeared in the market. Researchers are enticed towards the domain because of its increasing opportunities and challenges. IoT has so much to

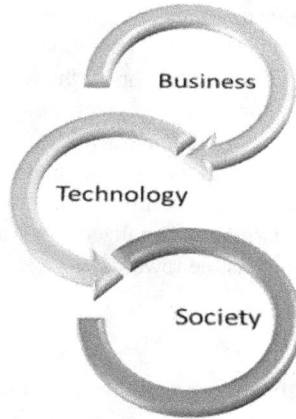

FIGURE 3.10 IoT and society.

contribute to society through businesses using the technology, as depicted in Figure 3.10. In this section, let us address a few working issues that arise with this technology.

3.7.1 COMPONENT CONVERGENCE CHALLENGES (CCC)

IoT and big data are based on four components [7]: people, data, processes, and devices/things, as depicted in Figure 3.11. People are using an enormous amount of data in their daily routines, which has to be collected using a device and processed to gain an insight from the data. The integration of these components is a big

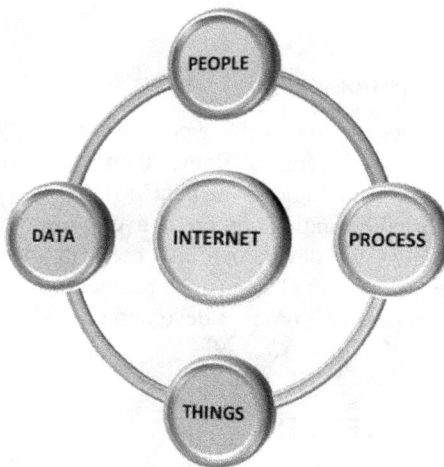

FIGURE 3.11 IoT and big data components.

challenge because of its heterogeneous nature. People are different because their businesses are different, hence the data will be different. The process has to be different because the data is different. Devices are different as the people, processes, and data are different. This interlinked technological chain obviously has challenges when dealing with the type of data it is processing in real-time. Human ability has increased because of a greater number of connected, controlled, and monitored objects. Very complex real-time tasks are easily possible, which has significantly reduced the processing time. The data has to be collected considering the Three Vs, so that the decision and prediction given by the process will always be reliable. The device should be capable of adhering to the people, process, and data as it is the vital part of data generation.

3.7.2 EMBEDDED NETWORK CHALLENGES (ENC)

Numerous protocols are used to collect data from the IoT device. The devices/things can be connected to the Internet through any medium of network, such as Bluetooth, Wi-Fi, or a wired/wireless connection, which is based on their reach, for example, a BAN (Body Area Network), PAN (Personal Area Network), CAN (Campus Area Network), or HAN (Home Area Network). The back-end services are working on the data received through these protocols. With this architecture, there are numerous threats to connectivity. It can be a hardware failure, software failure, or connectivity failure. Building a robust IoT system, in spite of these possible failures, is a challenge and it is always appreciated by and enticing to technology experts and researchers.

3.7.3 ANALYTICS AND APPLICATION CHALLENGES (AAC)

As the IoT is working with big data, the process of data collection varies in variety, volume, and veracity. The analytical challenges behind these Vs are numerous. The data structure maintained for the storage and the convergence of data with a variety of formats like structured, unstructured, and semi-structured have to be analysed and predicted before algorithmic development.

There is a necessity to balance the speed and scale of data. Moving Tbs of data over a Gbs network would be imprudent for real-time analytics. Therefore, the scale of conduct of algorithmic analytics has to be envisioned considering the challenges shown in Figure 3.12. In order to remain one step ahead, people prefer to be equipped beforehand for a problem rather than to react in real-time [8]. This has given rise to the largest-scale analytics called artificial intelligence (AI) with IoT.

3.7.4 ETHICAL AND SECURITY CHALLENGES (ESC)

The IoT has gained the attention of business companies, government initiatives, and people to meet societal and individual needs. This situation can create vulnerabilities if the user is not aware of the sensitivity of the data, information, or knowledge shared via the IoT pathway, which is prone to malicious attacks. Ethics often lag behind technological innovation [9].

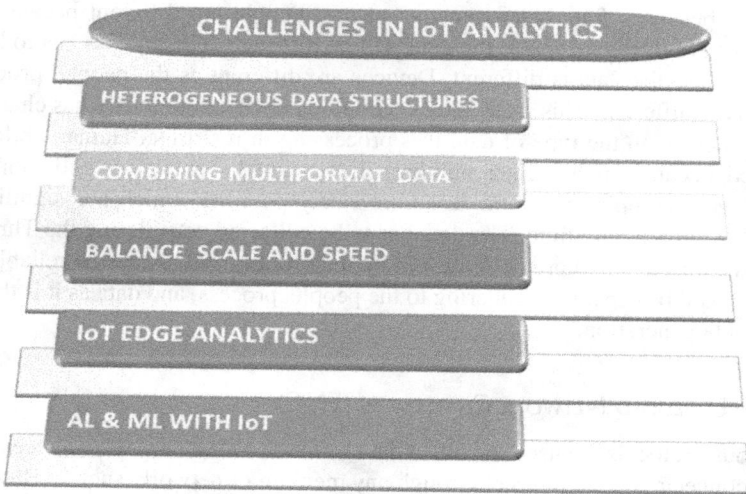

FIGURE 3.12 IoT analytical challenges.

The number of devices connected to the Internet is increasing day by day. Automated secured device identity authentication is becoming obligatory. Next, the communication between the device and the cloud should be secured. Using standard encryption via a network, isolated device networks are considered good practice for sending private and confidential data over the IoT. This will ensure data privacy and integrity among users. Despite considerable efforts, security vulnerabilities and breaches are foreseeable. The IoT development should construct strategies for detecting vulnerabilities, breaches, and anomalies by monitoring the network through effective penetration testing and ethical hacking. The enhanced use of intelligent security analytics in the IoT system will augment the trust of the people in the IoT arena.

IoT devices have become cost-effective nowadays as there is a fall in the prices of sensors. This makes consumers buy more items connected to the IoT. As IoT devices and IoT consumers are increasing day by day, our lives are filled with numerous IoT devices around us. Though technology helps us in our daily chores, it has to be cautiously used to feel the real sense of IoT, as the more digital it becomes, the more vulnerable it is. Figure 3.13 depicts various security challenges in IoT implementation.

3.7.5 IoT Adoption Challenges (IAC)

The IoT has revolutionized the industrial sector through many successful projects. The growth in sensor technology in terms of range, sensitivity, resolution, speed, coverage, and cost-effectiveness has led to such success. Despite serious efforts to succeed, the room for IoT adoptive challenges is inevitable. Data security concerns, lack of funds for IoT products, interoperability challenges, integration with the legacy system, complex IoT architecture, standards, and many more will add up to the

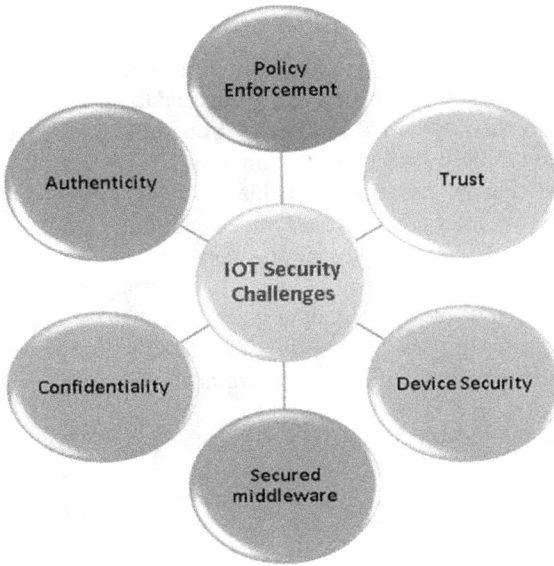

FIGURE 3.13 IoT security challenges.

IoT adoption issues as shown in Table 3.5. The quality of IoT technology can be distinguished when the product is made as robust as possible by bearing in mind the challenges discussed above and making the user understand the necessity of collaboration and interoperability between devices and networks.

To conclude, as the number of connected devices grows, the risk of data hacking also grows. Compliance standards are desirable to ensure all IoT device manufacturers are constructing devices that are protected by architecture for the artefact's complete lifespan. IoT security is a technology that monitors the physical and logical part of the system of connected devices in an IoT network. The state-of-the-art efforts in IoT security have been discussed as they are the important measures to be taken to protect the IoT ecosystem. IoT features and requirements vary from domain to domain. Therefore, any organization using the IoT should follow the IoT measures and ethical framework with multi-level encryption for secured data transfer.

TABLE 3.5
IoT adoptive challenges

Data security concerns
Insufficient funds for IoT projects
Ecosystem complexity
Legacy system integration
Challenges in interoperability
Connectivity and bandwidth challenges
Lack of skilled resources
Adherence to industry regulations

3.8   CONCLUSION

The potential of the IoT to enhance reality and its prospects for the future are inevitable. The hugely successful expansion of big data and the IoT is drastically influencing almost all the technologies and businesses around the globe. We can understand the role of big data in collecting massive amounts of data from the innumerable of devices. This data helps in the decision-making process following regression analysis and analytics. This continuous process may sometimes face issues related to data collection, data processing, and data security. These issues can be resolved with sensible use of intelligent IoT architecture and intelligent algorithmic system integration. The IoT and big data blend together to make us realize the need for IoT systems. To conclude, IoT and big data are two different domains but when combined they amaze and mesmerize people with their performance.

> "If you think that the internet has changed your life, think again. The Internet of Things is about to change it all over again"

> – Brendan O Brien [10].

REFERENCES

[1] A. A. Adewuyi, H. Cheng, Q. Shi, J. Cao, A. MacDermott, and X. Wang, "CTRUST: a dynamic trust model for collaborative applications in the internet of things," *IEEE Internet Things Journal*, vol. 6, no. 3, pp. 5432–5445, June 2019, doi: 10.1109/JIOT.2019.2902022.

[2] "What is big data? Behind the buzzword." [Online]. Available: https://www.swc.com/blog/business-intelligence/behind-buzzword-big-data. Accessed December 29, 2019.

[3] "Data never sleeps 5.0 I Domo." [Online]. Available: https://www.domo.com/learn/data-never-sleeps-5. Accessed December 30, 2019.

[4] M. Rakusen, *"Understanding Data Streams in IoT Contents,"* 2013.

[5] *"Five Stages of IoT,"* Bsquare Corporation. 2016.

[6] S. Balakrishna and M. Thirumaran, "Semantics and clustering techniques for IoT sensor data analysis: a comprehensive survey," in: Peng, S.L., Pal, S., and Huang, L. (Eds), *Principles of Internet of Things (IoT) Ecosystem: Insight Paradigm.* Intelligent Systems Reference Library, vol. 174. Springer, Cham, 2020, pp. 103–125.

[7] Z. Alansari et al., "Challenges of Internet of Things and big data integration," *Lect. Notes Inst. Comput. Sci. Soc. Telecommun. Eng. LNICST*, vol. 200, no. August, pp. 47–55, 2018, doi: 10.1007/978-3-319-95450-9_4.

[8] Ahmed Banafa "Five challenges to IoT analytics success – OpenMind." [Online]. Available: https://www.bbvaopenmind.com/en/technology/digital-world/five-challenges-to-iot-analytics-success/. Accessed February 8, 2020.

[9] F. Allhoff, "Risk, precaution, and emerging technologies," *Studies in Ethics, Law, and Society*, vol. 3, no. 2, 2009, doi: 10.2202/1941-6008.1078.

[10] 19 astonishing quotes about the internet of things everyone should read. https://www.forbes.com/sites/bernardmarr/2018/09/12/19-astonishing-quotes-about-the-internet-of-things-everyone-should-read/#6485c909e1db, Accessed August 4, 2020.

[11] "'The digital revolution is over': Nicholas Negroponte in 1998.". https://escherman.com/2008/03/12/the-digital-revolution-is-over-nicholas-negroponte-in-1998/. Accessed October 7, 2020.

4 Compulsion for Cyber Intelligence for Rail Analytics in IoRNT

Nalli Vinaya Kumari,
Sri Satya Sai University of Technology & Medical Sciences, India

G. S. Pradeep Ghantasala*, and
Mallareddy Institute of Technology and Science, India

M. Arvindhan
Galgotias University, India

4.1 INTRODUCTION

In computer science, just as natural intelligence is displayed by human beings, artificial intelligence (AI) is the intelligence that robots display. Within AI, the area is defined by leading algorithms that understand the world and take actions that maximize their chance to achieve their objectives. The term "artificial intelligence" can be used colloquially often to describe robots (or computers) that imitate "cognitive" features, such as "thinking" and "problem-resolution" associated with human cognition. With robots becoming ever more intelligent, "thinking" functions are often taken out of the AI context, a phenomenon known as the AI effect. A point in Tesler's Theorem states that "AI is something that hasn't been achieved yet" [1]. For example, the detection of optical properties is mostly omitted from AI, because they are a standard technology. Data mining is the mechanism by which, at the crossroads of computer science, analyses, and information systems, large data collections of methods are identified. Data mining is interdisciplinary computer science nd analytics related to data processing (with logical methods), giving it a coherent framework for further use. In addition to the actual computational stages and the aspects of database and data management, data processing and pre-processing, ideal and estimation criteria, value measures, complexity evaluation, post-processing discovered systems, simulation, web updates, data mining is a web-based or KDD database discovery application. Data mining is a research stage of the data processing process. Practical machine learning (ML) software and methods for Java were first known as "The words 'data analytics' and 'analytic' (large-scale) – or artificial intelligence and 'machine learning' – [were] used for scientific machine learning". The word 'data mining' was used for advertising only. The actual data mining task is to examine large amounts of previously unknown or curious phenomena like data record classes (cluster analyses), odd data (anomaly detection), and associations (association laws, serial sequence mining). Data mining is the true function of semi-automatic or automated computing. It usually means that methods such as temporal indexes are used in the search. Such models can then be used as a way to view input data, for instance in the field of ML and prediction processing, for further analysis. For instance, many classes of knowledge can be represented in the data mining process, which can be used to generate more accurate outcomes through a decision-making support system. No data collection, compilation, measurement, or monitoring is part of the process of data mining. In addition, further steps in the general KDD method are included. The distinction between data analyses and data mining is that data analyses are used to evaluate data set structures and theory, for example, analysing the output of a marketing campaign, irrespective of the quantities of data. In contrast, data mining uses machines and mathematical models to identify clandestine and latent anomalies in a wide range of data. Information dragging, information digging, and the snooping of data refer to the use of data mining techniques to sample segments of a larger population data collection that is (or may be) too small to be used for the legitimacy of any correlations observed. Nevertheless, these approaches can be used when new theories are formed to check large populations of data. In the 1960s, researchers and analysts used terminology such as data mining or dragging as the wrong way to analyse data without a predetermined theory, which they considered a bad idea. Economist Michaël Lovell also used the term "data mining" in an article published in the 1983 *Journal of Economic Studies*; the word data mining appeared in the college in 1990, often with positive connotations [2]. The term "information mining™" was used in the 1980s, but because the San Diego-based HNC company regulated it, it is also known as "masking". It varies from "experimentation" (positively) to "fishing" or "denial". Many terminologies used include the archaeology of records, data collection, material discovery, knowledge retrieval, and so on. Nevertheless, data mining became more popular in industry and the media. Big data is a field where data sets that remain large or complicated to use and where outdated data processing instruments are discussed, gathered endlessly, or else treated. Big data's recent usage applies to the practice of predictive analysis, consumer experience analysis, or some additional progressive methods of data analysis that collect data. The data sets will classify the new links in industry patterns, disease prevention, violent crime prevention, and the like in the areas of web searches, fintech, digital computing and company IT, academics, business leaders, and practitioners of medicine. There is no question that there are large numbers available today, however, that's not the greatest important features of the different environment.

4.2 COMPUTER-BASED INTELLIGENCE

While artificial intelligence and computational intelligence (CI) pursue a similar long-term goal − to achieve general intelligence, which is the intelligence of a computer capable of performing any intelligent function that a human being can perform − there is a clear difference between them.

Machine intelligence is a subcategory of AI, according to Bezdek [3]. There are two forms of machine intelligence: AI based on hard computation and soft computation, which can be reviewed in some situations. Fast computing techniques work

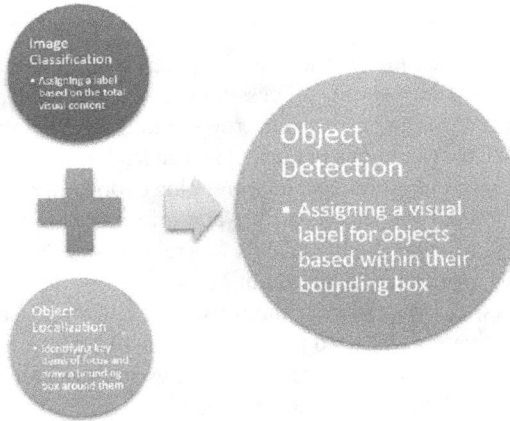

FIGURE 4.1   Classical intelligence for identification.

according to binary logic based only on the two principles on which the current computers are based (Boolean true or false, 0 or 1). One difficulty with the philosophy is that the exact terms 0 and 1 of our natural language are not always appropriate. Weak, thuggish-logical computation methods can be helpful here [4]. This theory is one of the most exclusive features of CI, far closer to how the human mind operates by applying details to partial truths (crisp and thuggish systems). Crisp and fluid systems adopt the same concepts of static and discrete logic. Designing computer-based instruction (CBI) using AI approaches can be viewed (see Figure 4.1) from the perspective of management systems. Instructional design variables resulting from the Minnesota Adaptive Instructional System (MAIS) programmatic research effort have been reviewed. An essential part of MAIS is its AI principles. Tutoring is tailored iteratively to every student's needs at the time. The six architecture variables of the MAIS have been tested following a brief summary of AI approaches ideal for the intellectual CBI. The extensive collection of empirical research results, which assists the program capabilities of the MAIS, is a unique feature.

4.2.1 CYBER THREAT INTELLIGENCE

This essay is the first of a series of updates released on Cyber Risk Intelligence and Intelligence Analyses (I&AWG) from the Multi-State Information Sharing and Analysis Centre. From this blog, we will research the intelligence of the cyber threat and analyse its importance to the members of the MS-ISAC, the challenges inherent in creating cyber risk intelligence, and various components of information, such as Words of Estimated Probability.

Cyber threat intelligence is information on cyber threats that is compiled, assessed, and interpreted with systematic and structured analysis methods by those with specialized know-how. Like all intelligence, cyber threat information allows users to identify threats and opportunities and thus reduce uncertainty, adding value to cyber threat results. To yield precise, appropriate, and meaningful information, analysts must process immense amounts of data and consider disappointments.

4.2.2 Big Data and Analytics

Big data is an area in which information sets that are considered too large or complicated to handle using traditional data processing software are analysed, systematized, or otherwise processed. Some cases (more rows) have higher statistical power, and low-complexity data (more columns or attributes) can lead to sophisticated false detection rates [5]. Data capture, data storage, interpretation of documents, scans, exchanges, transfers, displays, searches, updates and privacy information, and database sources are significant challenges. Originally, big data was related to three key concepts: quantity, variety, and time. We cannot analyse but observe and watch what happens as we handle large numbers. Big data also contains estimates with measurements that surpass the standard software's ability to process them due to expense and lack of time. Present-day use of big data includes predictive analysis, user behaviour analysis, and specific other advanced methods of data analysis to extract important data and to set the scope for accurate data. The data collection research will identify new ways to "spot market patterns, [avoid disease], [fight] violence, etc." (see Figure 4.2).

With the volumes of data currently available, there is little uncertainty, but that is not an essential characteristic of this new data environment. In the fields of Internet searches, fintech, digital computing, and business knowledge, physicists, business managers, medical practitioners, marketers, and governments alike regularly experience difficulties with vast sets of data. Scientists face restrictions in e-science, such as meteorology, genomics [6], connectomics, dynamic models of physics, biological research, and environmental research [4]. Data sets are expanding rapidly to a certain point, for they are increasingly accumulated through the inexpensive and abundant IoT, such as mobile devices, remote sensing, software diaries, cameras, microphones, radio frequency identification (RFID) readers, and wireless sensor networks (WSNs) [7]. According to an International Data Corporation (IDC) survey, the world volume of data will exponentially increase between 2013 and 2020 from 4.4 to 44 zettabytes. In 2025, the IDC predicts that 163 zettabytes of data will be available [8]. One question for big firms is who will own large-scale data projects affecting the entire organization [7]. The term "big data" has been used since the 1990s, with John Mashey popularizing it [9, 10]. Big data typically includes data sets extending beyond the capability of software tools currently available for the gathering, storage, and analysis of data [11]. The "size" of large-scale data is continually changing from a few

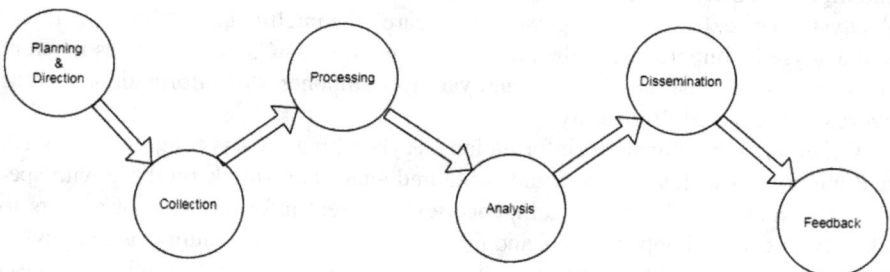

FIGURE 4.2 Life cycle of cyber threat intelligence.

hundred terabytes up to plenty of zettabytes of data in 2012 [12]. Big data requires a set of new types of convergence systems and techniques to reveal information on large, complicated, and mass-scale data sets [13].

4.3 ANALYTICS TYPES: DESCRIPTIVE, PRESCRIPTIVE, AND PREDICTIVE

The Big Data Revolution gave rise to many different data analysis types and stages. Boardrooms across companies are boosting data analytics – providing wide-ranging business success solutions to companies. However, what do these say to companies? It is only the right data that enables companies successfully using big data to increase economic success. Big data modelling seeks primarily to help businesses to determine faster and better results (see Figure 4.3).

Comprehensive data analytics cannot be accepted as a single-size-fits-all approach. Besides, the best data scientist or data analyst can define the kind of research that the company can make use of. The three major analytical styles, descriptive, predictive, and prescriptive, are interrelated instruments that help businesses leverage their results. Each of these theoretical methods offers a different point of view. This chapter explores the three different types of analysis – descriptive analysis, predictive analysis, and prescriptive analysis – to discover how every kind of analysis improves a company's operational efficiency (see Figure 4.4).

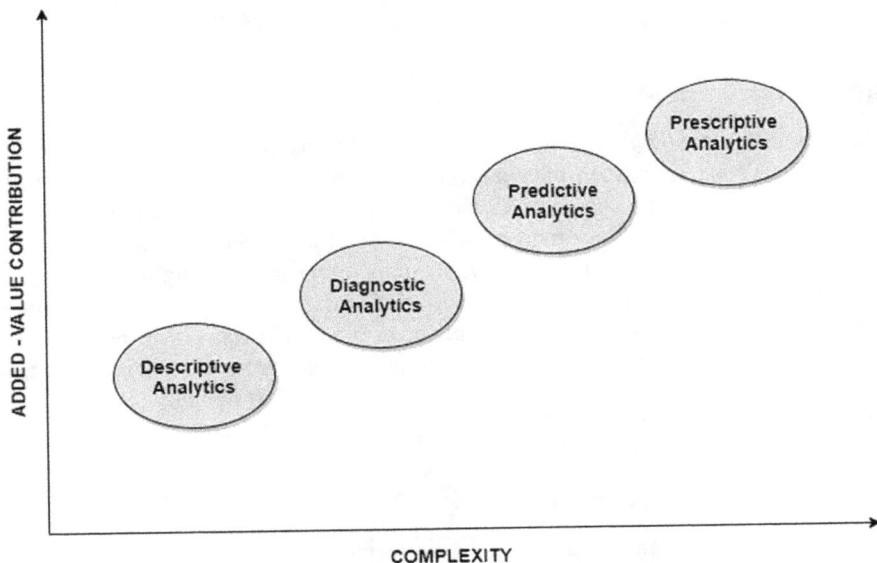

FIGURE 4.3   Added-value contribution vs complexity.

Descriptive Analytics	Predictive Analytics	Perspective Analytics

What Has happened?	What could happen in the future based on previous trends and patterns?	What should a business do?

FIGURE 4.4 Illustration of descriptive analytics, predictive analytics, and prescriptive analytics.

4.4 UNDERSTANDING PREDICTIVE AND CONCISE ANALYSIS

To help to find her victims, a lioness hired a data scientist (a fox). The fox was able to access a rich data store full of forest records, covering its species, and activities that were taking place in the jungle. On the first day, the fox delivered a summary to the lioness, which described where her prey had been found in the last six months. This is a descriptive analytics example. The fox then calculated the probability of finding a single beast, using sophisticated ML technologies, at a particular place and time. This is predictive analytics. The lioness established paths in the forest to reduce efforts to find her prey. This is a case in point regarding optimization. Finally, the fox created traps on the basis of the above models so that the prize was automatically picked up at several locations in the jungle. This is robotics automation.

4.4.1 ANALYTICAL METHODS

Big data analytics allows a company to identify the customers' needs and expectations, and firms can broaden their customer base and preserve existing customers through tailored and specific deals for their products or services. The IDC forecasts that the big data and analytics markets will rise to a total of $41.5 billion by the end of 2018 at a 26.4% CAGR. Thanks to varied technologies, including intelligent grid controlling, sentiment exploration, fraud-finding, customized deals, traffic monitoring, and so on, through a variety of sectors, the big data industry is growing exponentially. With the processing of big data by companies, automation is the next important step. Most businesses do not know where to begin, what sort of analytics will facilitate business growth, or what these various analytical styles mean.

4.4.2 DESCRIPTIVE ANALYTICS

Today, 90% of companies use the most basic method of concise analytics. This answers the question, "What happened?" This type of analysis involves analysing real-time and historical data for insight into how the future can be addressed.

The main goal of analytical research is to determine whether notable success or failure has been accomplished in the past. The "present" here applies to any moment an event took place and might be a month before or only a minute before. A company learns from previous behaviour and understands how it will affect future results. Descriptive analysis is used when a company needs to grasp and explain different aspects of the overall performance of the company. Descriptive analytics is focused on regular aggregate database functions that only require simple school mathematics expertise. Descriptive analysis is the core of the social critique. Many groupings are listed based on the pure numbers of some instances. There are only case registers, the number of followers, comments, tweets, and supporters. For social analysis, these metrics are the products of basic mathematical operations, such as average response times, average response percentages, percentage index, number of page views, and so on (see Figure 4.5). The results obtained by a company using Google Analytics software from the web server are the best illustration of informative analytics. The findings help us to understand what has taken place in the past and to determine whether or not a marketing camp has been successful, based on fundamental metrics such as page views.

4.4.3 PREDICTIVE ANALYTICS

Predictive analytics is the next step in data reduction. Analysis of past data habits and developments will tell what could occur in the organization's future. It leads to real business goals, efficient preparation, and aspirations of elimination. Predictive analytics are used by businesses to research the data and to find the answers to the search. Predictive analytics attempts to forecast the possibility of a potential event through several mathematical and ML algorithms, but prediction accuracy is not 100%, because it is dependent on probabilities. Algorithms use the data and fill the missing data with the best estimates to make predictions. The data is collected using CRM, POS, ERP, and HR historical data to evaluate data patterns and relations among different data set variables. Companies should focus on recruiting strategic experts, selecting ML algorithms for quantitative modelling, and implementing a successful organizational strategy.

Predictive analytics can be defined as, "What's next if?"

- Study Root Cause – Why was it?
- Data Mining – Correlated data detection
- What if current trends continue? What is the forecast?
- Monte Carlo Simulation – What might happen?
- Pattern identifying and alerting – When action to fix a pattern should be introduced.

The most common type of predictive analytics is sentiment analysis. The learning model uses everyday language, and the output is a score that helps to decide the positive, negative, or neutral feeling of the experiment. Organizations such as Target, eBay, and other retailers employ forecast analytics to identify trends in customer buying habits, make inventory projections and supply forecasts, identify items that

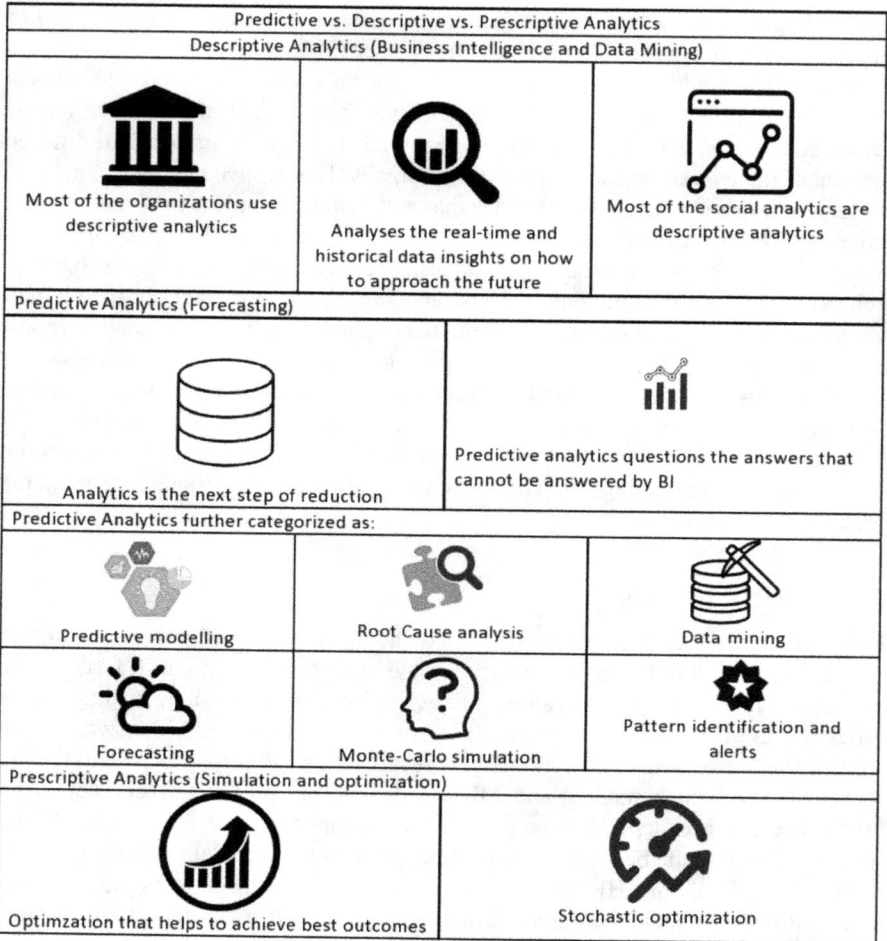

Predictive vs. Descriptive vs. Prescriptive Analytics		
Descriptive Analytics (Business Intelligence and Data Mining)		
Most of the organizations use descriptive analytics	Analyses the real-time and historical data insights on how to approach the future	Most of the social analytics are descriptive analytics
Predictive Analytics (Forecasting)		
Analytics is the next step of reduction	Predictive analytics questions the answers that cannot be answered by BI	
Predictive Analytics further categorized as:		
Predictive modelling	Root Cause analysis	Data mining
Forecasting	Monte-Carlo simulation	Pattern identification and alerts
Prescriptive Analytics (Simulation and optimization)		
Optimzation that helps to achieve best outcomes	Stochastic optimization	

FIGURE 4.5 Thomas Jefferson on descriptive vs predictive vs prescriptive analytics.

consumers are likely to buy in tandem so that they can make personalized recommendations, and estimate the quarterly or yearly sales volume. The most useful indicator of predictive analytics is the glory nick. A glory nick allows fiscal institutes to determine whether a borrower is likely to pay credit bills on time.

4.4.4 PRESCRIPTIVE ANALYTICS

Big data may not be an accurate crystal ball to predict the exact number of winning numbers. Still, it can illustrate problems and help an organization understand why those problems occurred. Busy companies can make solutions for market problems that lead to successes and insights using data-based and data-driven variables.

Prescriptive analytics is the next step to add a flavour to influence the future. Prescriptive analytics inform on potential results and result in decisions that optimize essential business metrics. Simulation is ultimately used to say, "What should an organization do? Prescriptive analysis is a state-of-the-art, optimization-based approach that helps achieve the best possible results. Stochastic optimization is used to understand how the best results can be made to identify data uncertainties for better decision-making. In order to simulate the future, the scenario design, based on different assumptions, can be done, which blends with different strategies of optimization. Prescriptive research evaluates many possible actions and recommends behaviour based on the results of a single sample descriptive and predictive analysis. A required report is a combination of data and different business rules. Data may be internal and external." Prescriptive analysis can be used to describe how best to achieve outcomes and forecast data complexity to decide more what will become of them.

Organizations gather descriptive data and link it through predictive analysis to other consumer user-compliance and web server results. While some companies stay as they are, corporations will expect market growth in the future. Predictive modelling offers better advice and future-oriented answers to questions not answerable by business intelligence (BI). For example, by using drug analytics, Aurora Health Care System lowers re-admission rates by 10%, saving $6 million per year. Prescriptive technology can be used in the area of healthcare to optimize drug development, finding the right candidates for clinical trials, and so on. As more and more businesses are recognizing the competitive advantage of big data, they must select the right kind of data analytic tools that maximize return on investment (ROI), minimize running costs, and improve the quality of service.

4.5 RAILWAY NETWORKS

India's Indian Railways (IR) is the Indian Ministry of Railways' primary railway system. As of March 2017, the company operates the fourth leading railway network in the domain with a road length of 67,368 km. Roughly 50% of roads were operated by electrical 25 kW 50 Hz AC and 33% by double or multitasking [6, 14]. In March 2018, the number of passengers carried by IR was given as 8.26 billion, while freight volume was 1.16 billion tons. IR runs 20,000 commuter trains a day, both on long-distance and regional lines, through 7,349 stations in India [14]. The bulk of the luxury passenger trains, such as Rajdhani, operate Shatabdi Exp at a peak speed of between 140–150 km/h (87–93 mph) and Gatiman Express between New Delhi and Jhansi at a maximum top speed of 160 km/h (99 mph). The average rate is 50.6 km/h (31.4 mph). Express trains are the most frequent. The average freight train speed is around 25 km/h (15.5 mph) [4]. Based on axle load and specific container speed at a top speed of 100 km/h (62 mph), a typical freight train's speed is between 60 and 75 km/h (37.2 and 46.6 mph). In March 2017, the IR's rolling stock consisted of 277,987 vehicles, 70,937 kerbside trains, and 11,452 steam engine [14]. In several Indian locations, IR operates locomotive and coach services. In March 2016, it had 1.30 million workers and was the eighth largest employer in the world [3].

4.5.1 Industry Pan Rail Directions

The required analysis is a combination of data and different business rules [15]. Data may be both internal (inside) and external (such as data from social media) for prescriptive analyses. Business rules are preferences, ethical practices, limits, and other restrictions. Mathematical models include natural language processing, ML, statistics, operational research, and so on. The railway network is similarly ideal for long-distance travel and transporting commodities, distant from actuality efficient and cost-effective manner of transference and conveyance. IR was the favoured exporter of saloons in the country with exports of sedans growing 16% in 2017–2018. The Government of India has concentrated on spending on the railways by partaking in investor-friendly procedures. It has stimulated hastily to allow Foreign Direct Investment (FDI) in railways to increase support for consignments and also high-speed trains. Many internal and international concerns want to capitalize on Indian rail ventures as well.

4.5.1.1 Marketplace Size

Revenue from IR rose to USD 27.13 billion in FY19 from 6.20% for CAGR over FY08–FY19. Passenger revenue was increased to USD 7.55 billion during 2018–2019P at a CAGR of 6.43% in FY07–FY19. Freight revenue increased by 4.30% to USD 18.20 billion in 2018–2019 in the CAGR for FY08–FY19.

4.5.1.2 Investment/Evolution

Between April 2000 and June 2019, railway-related components obtained USD 977. 24 million from FDI.

Here are specifics of the first reserve and innovations in the railway zone in India:

- A pilot venture was initiated in November 2019 to investigate the possibility of e-tail players in the use of a railways parcel service.
- The maintenance deal with Madhepura Electric Locomotive Pvt Ltd. (MELPL) was signed by IR in November 2019. A joint agreement was made between IR and Alstom, a French-based company, for the supply of 800 electrical freight services steam engine and their related maintenance.
- One-Touch ATVM was launched in October 2019 for fast tickets at 42 central railway suburban stations.
- The eastern highway in Uttar Pradesh, Khurja–Bhadan, was opened officially for traffic on October 2, 2019.
- In July 2019, the longest electrified tunnel was built, connecting Cherlopalli and Rapuru stations.
- An aggregate of 24,493 bio-toilets was fitted in trains between April and August 2019.
- In May 2018, to provide berths on alternative trains to passengers on waiting lists, IRCTC launched the Alternate Train Accommodation Scheme (ATAS).
- Alstom, a company based in France, reported in December 2018 that it plans to increase its coach production in Sri City from 20 to 24 vehicles a month.

4.5.1.3   Government Initiatives

Some new government programmes are as follows:

- Indian Railways announced in February 2019 the introduction of food packages with QR codes and live kitchen service.
- On the Delhi–Mumbai and Delhi–Howrah roads, the velocity will be increased to 160 km/h (99.4 mph) by 2022–2023. Passenger trains are allowed to increase their average speed by 60%.
- Under the 2019–2020 budget of the Union, Rs 160,176 crore (USD 22.91 billion) was allotted by the Government of India to the Ministry of Railways.
- The main two merchandise passageways, Eastern Freights Passageway from Ludhiana to Dankuni (1,856 km) and West Freights Passageway from Dadri to Jawaharlal Nehru Port (1,504 km), are already being built at a budget of Rs 81,000 crore (USD 11.59 billion). The Indian Freight Corridor (DFCCIL) has already been built.
- IR plans to develop a new railway export strategy in November 2018.
- The Government of India will create a "National Rail Strategy" to allow it to merge the existing rail network into further transport means and construct a multi-modal transport system.
- To ensure modern and consistent processes and procedures, a "Novel Online Registration System for Vendors" was proposed by Research Designs and Norms Organization (RDSO), a research arm of IR.

4.5.1.4   Road Ahead

- There is rising interest in the Indian railway network (see Figure 4.6). The Indian railway market, with 10% of the global marketplace, will be the third largest in the next five years. According to Piyush Goyal, Railways and Coal Minister of the Union, IR, alone among the country's principal proprietors, will create 1 million jobs.
- The IR's goal is to increase its carriage traffic from 1.1 billion tons in 2017 to 3.3 billion tons by 2030.

Advantage	Growing demand	Increasing urbanisation and rising incomes are driving growth
	Attractive Opportunities	Freight traffic is set to increasable significantly due to rising investments and private sector participation
	Higher Investments	As per the budget As per the expected investments
	Policy Support	Providing and supporting policies roles

FIGURE 4.6   Rail advancements.

- The freight traffic through the designated freight corridors is expected to increase from 140 MT in 2016–2017 in the CAGR from 5.4% to 182 MT in 2021–2022.

4.6 RAIL ANALYTICS

An excellent BI method designed to increase productivity in the growth of the train network and simplify the transportation processes is Prognoz Railway Company Analytics [12]. This provides reliable information on the present and future state of road infrastructure, reduces spending on track maintenance operations, increases expenditure performance in construction projects, and handles electronic freight traffic. The system offers comprehensive data collection, consolidation, and updates for tracking infrastructure and objects, including GPS, descriptions, and image coordinates. The system's broad versatility helps to customize routes based on criteria such as size, distance, extra freight handling, off-scale cargo transport, and so on. Figure 4.7 indicates the nature of the rail system.

The programme also helps to forecast the amounts of operations needed for sustaining and renewing infrastructures by making the budget process simpler and eliminating adverse effects of underfunding. Prognoz Railway Company Analytics has been designed for railway company executives and shareholders, financial and economic division consultants, researchers, designers of planning operations, and marketing experts. Big data analytics (BDA) has attracted increasing attention from analysts, researchers, and train and engineering professionals. This calls for a study of the recent development of research in this area. A comprehensive review by Mayring [16] of the latest developments of big data in the area of rail and transport will be performed by this chapter. The survey encompasses three fields of rail transport in which BDA is applied: operations, maintenance, and safety. The level of big data analysis, significant data model forms, and variability of extensive data practices must also remain examined and evaluated.

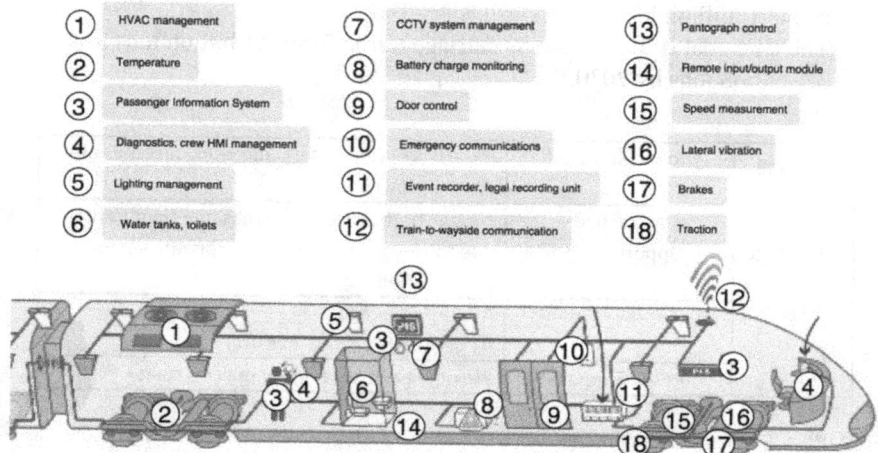

① HVAC management	⑦ CCTV system management	⑬ Pantograph control
② Temperature	⑧ Battery charge monitoring	⑭ Remote input/output module
③ Passenger Information System	⑨ Door control	⑮ Speed measurement
④ Diagnostics, crew HMI management	⑩ Emergency communications	⑯ Lateral vibration
⑤ Lighting management	⑪ Event recorder, legal recording unit	⑰ Brakes
⑥ Water tanks, toilets	⑫ Train-to-wayside communication	⑱ Traction

FIGURE 4.7   Illustration of the nature of the rail system.

4.7 INTERNET OF RAIL NETWORK THINGS

The Internet of Things (IoT) is just one term for a sophisticated computer, network, and software networking, which drives outside machine-to-machine (M2M) infrastructures. This requires a variety of standards, areas, and also programs. The IoT helps freight and commuter roads to gather and evaluate information from various channels and data streams with the aid of M2M analytics tools, "Big Data" processing, cloud computing, and other technologies, and then use this analytical approach to improve efficiency, control processes, and possibly deliver new services. The IoT thus presents railways and infrastructure suppliers with a world of opportunities and regional strategists it should tap into and engage with the new phenomenon in all the railway departments – from IT to shipping, electronics, and mechanics, via communications and signals (see Figure 4.8). "It isn't 'Create, and they're going to come', it's 'What's the result you want to drive?' There is no other way to be a global industrial technology supplier for air, mining, maritime fuel, stationary and forging

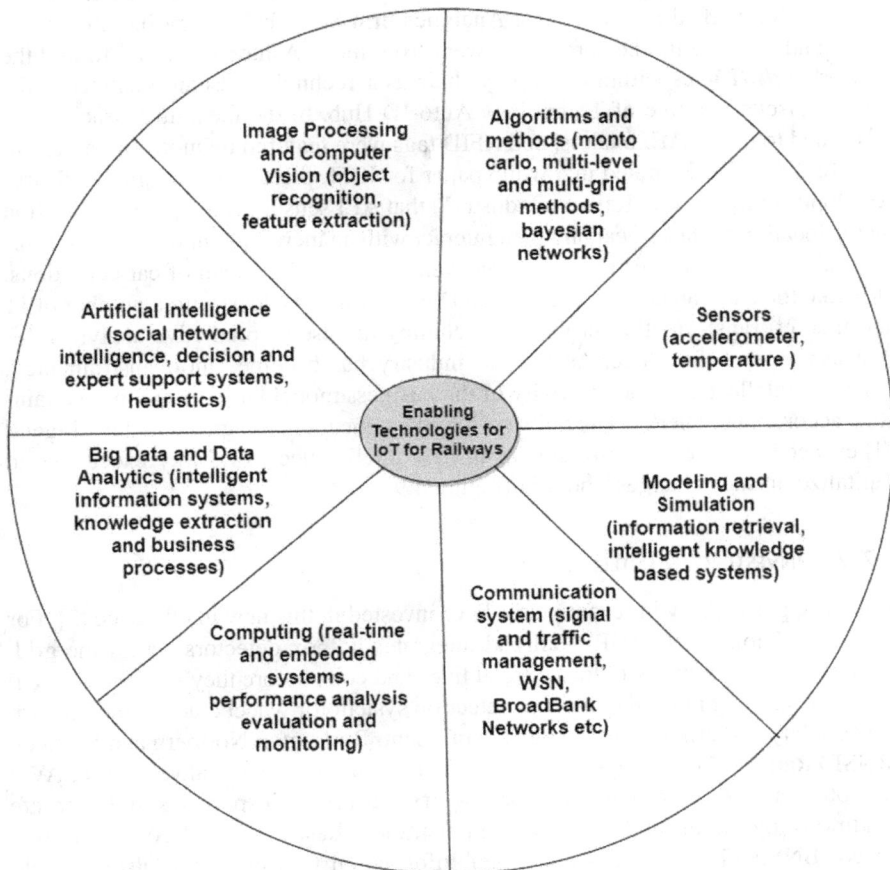

FIGURE 4.8   Enabling technologies for IoT for railways.

sectors," says Jamie Miller, President and CEO of GE Transportation. "You are working back into the data from there."

4.7.1 FROM APPLICATION ENABLING INTERFACE TO IOT

In railways, seeds were planted with application enabling interface (AEI) identifiers, the first of which was mounted in North America in 1989. In the United States, the seed data has seceded. A wayside detector sent a warning to the train crew, and the AEI readers updated a database of information and instructions. The tags were passive; the reader provided the power. In the mid-1990s, the general public were well on their way to untangling the Internet, but there was not precisely a pre-acceptance of it by the North American railroads. However, at Procter & Gamble (P&G), Brand Manager Kevin Ashton put RFID tags on products while going through the supply chain and connected them to the Internet. "First of all, it was just RFID," says Keith Dierkx, IBM's former rail industry pioneer and founder of the Former Rail Innovation Center. In 1999, Ashton coined the term "IoT", which he used in the title of a talk he made at P&G, according to conventional IoT wisdom. Ashton and others – including Dierkx, who worked for the Sensor Analytics firm Savi Technology Inc. in the late 1990s and stayed with the company – went from there. Ashton continues to find the networking/IoT gaps within the supply chain as a Technology Board member of the Massachusetts Institute of Technology Auto-ID Hub. In the meantime, track roads continued to install AEI readers, and RFID tags were featured by more and more rail cars. In 2009, Dierkx noted in a white paper for IBM, titled "The Smarter Railroad: An Opportunity for The Railway Industry", that AEI sensors could give information on the location of the rail car and then interact with an increasing number of monitoring and detection devices to provide an even more detailed panel of car conditions. "It's like the Russian nesting dolls," said Dierkx, who has dealt with a number of IT and market transformation problems, including the use of predictive analytics, for railways around the globe. As the rail industry has become "more instrumented, linked so intellectual", the creativity of the business model has become more attainable, according to Dierkx. The railways have become more integrated and intelligent. "They need to accelerate investment in new intelligence to railway executives to capitalize on such changes," he said (Figure 4.9).

4.7.2 INVESTING IN INTELLIGENCE

Railways, particularly in recent years, have invested in this new intelligence [3]. For example, Union Pacific (UP) Railroad integrates hotbox detectors across the grid, creating a single network to locate trend lines and coils before they malfunction. UP is also developing its own ultrasonic detection system for wheel crack, which in turn drives safety and efficiency to avoid a derailment. Burlington Northern and Santa Fe (BNSF) Railway Co. was approved last year to use unmanned aerial vehicles (UAVs) to capture a video after the crash, or to carry out bridge inspections under unsafe conditions, by the Federal Aviation Administration. Last year, the UAV permit was issued. BNSF Vice President and Chief Information Officer Joann Olsovsky told participants in the 2015 GE Summit Minds + Machinery in San Francisco, September

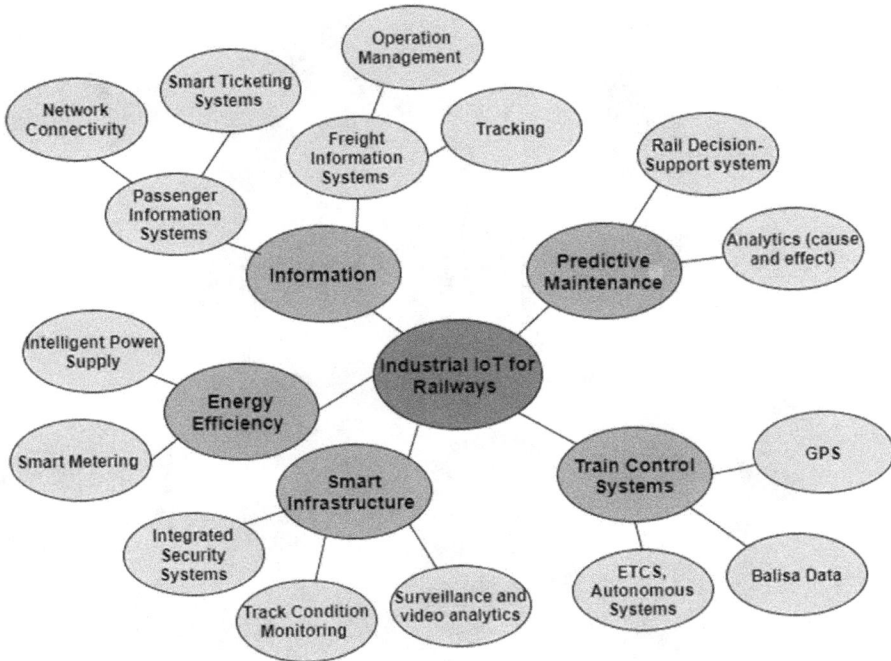

FIGURE 4.9 IoT for railways and its wide areas.

29–October, 1, 2015: "We are using video from these UAVs and hope that we can predict failures before they happen." But all major freight routes invested in remote and in real-time surveillance technology for railways and locomotives, track conditions, and component condition; shorter roads and regional authorities are also becoming increasingly important and the amount of passenger and transit railway companies. "All this must be synchronized," said Progressive Railroading Canadian National Vice President and Chief Information Officer Serge Leduc in 2013. In terms of incorporation, application vendors are more than pleased to help. As General Electric Chairman and Chief Executive Officer Jeff Immelt told Minds + Machines in 2017: "Our CIOs and IT professionals are the biggest drivers of production in our business." GE launched a marketing campaign in the autumn of 2011 to strengthen the position of the "digital industrial company". The aim was to increase profitability also for consumers. GE's Tier 4 series EvolutionTM locomotive was released in 2012 and tested in the field in 2013. It features over 200 onboard sensors that track health and performance, process over one billion orders per second, and provide about 10 GB of data per year. By 2013, GE has provided North American customers with more than 500 Tier 4 compliant locomotives. "These systems have been so incredibly caught," says Miller, a GE President and CEO who was elected GE SVP and CIO last September. As an example of that value capture, Miller cites the company software Trip Optimizer. The app creates an optimized service schedule based on information gathered about the features of a vehicle, mounted in the North Americas on 7,000

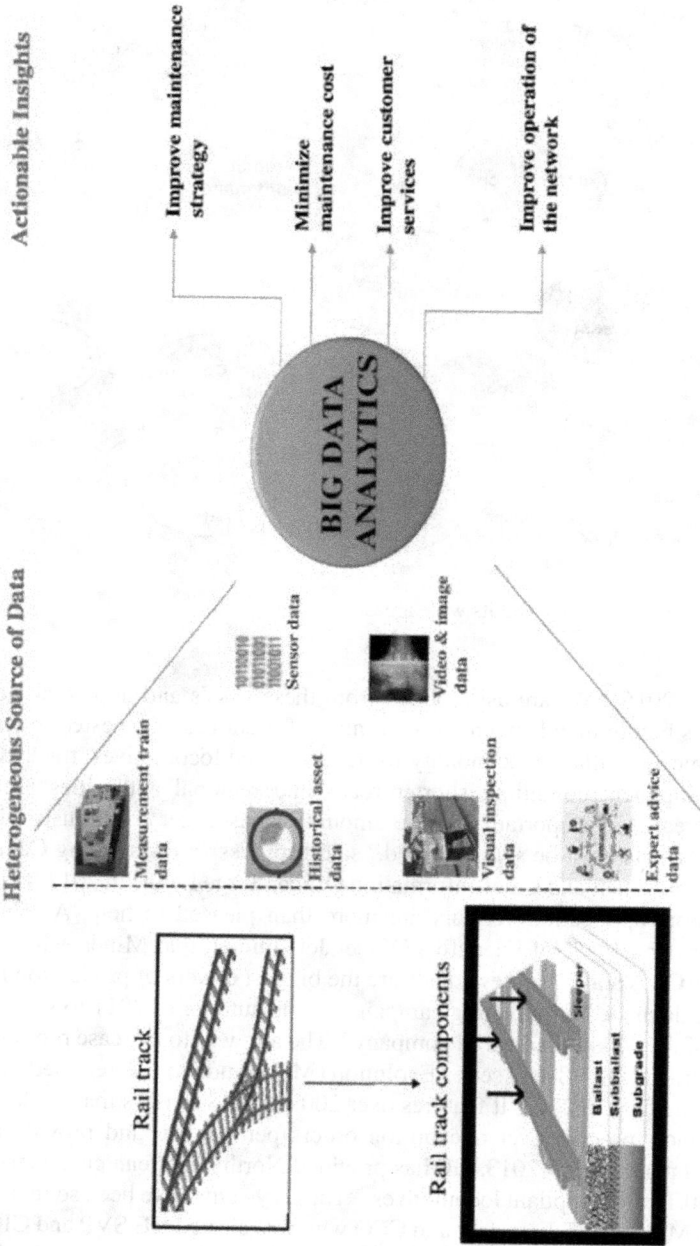

FIGURE 4.10 Big data development in the rail industry.

locomotives. Miller says that in 2015 consumers have saved USD 200 million in fuel costs. Seth Bodnar, GE Chief Digital Officer of Transportation, has also aimed to "unlock interest" for other GE technologies from the travel manager (crew management) to yard plane (hump yard optimization) to OASIS (operational test of intermodal yards). All of this is linked to GE's PredixCloudTM, a service platform. Predix collects and analyses computer data in a stable cloud environment, volume, rapidity, and range. Customers and other industrial companies will soon be available to handle data and software on a commercially viable basis. Technology companies are also serving commuter trains. In 2014, Cisco Systems introduced Connected Rail, a solution designed to upgrade aging rail networks, enhance security, reduce costs, and offer passengers an "enhanced connected experience," which includes the positive train control component in the integration of onboard, track, roadside, signalling and backing systems (see Figure 4.10).

4.8   BIG DATA IN RAIL INTELLIGENCE BASED ON CYBER THREAT

Tommaso Spanevello, Public Affairs Director at the Association of the European Rail Industry (UNIFE), has outlined the productive cyber security policy for the rail sector with the growth of big data and resulting rise in cyber threats and has underlined the need for more inter-industry coordination to benefit eventually from rail's performance.

The digitization and dissemination of web-based technology fundamentally affect the way companies, associations, communities, and individuals interact. During this age of the IoT, a growing network of items (such as laptops, tablets, and smartphones), records, processes, and people are linked to each other. When IoT slowly becomes the "Internet of All", such a permanent connection creates a fast-growing volume of data generated, exchanged, and processed. The importance of data collection, management, and sufficient processing by the rail supply industry is increasingly recognized and promises a profound transformation of the railway industry. We can gain practical insights into how to improve the results of the rail business by collecting and processing this data. The growth of big data and cyber threats in the digitalization sector and the expansion of web-based technology are changing the patterns of communication between enterprises, organizations, communities, and people.

In reality, we can gain practical lessons by gathering and analysing this data, which allow us to optimize the performance of rail operations. On the one hand, rail companies are provided with business intelligence to improve performance and automate their plans through active management and data processing; on the other hand, it would reveal massive amounts of confidential and personal data and enhance cyber threats. Also, technological progress and the use of extensive data redefine the security environment entirely as networks always become vulnerable to new kinds of risks. There is thus a great interconnection between the growth of big data in the rail segment and the development of cyber security. In its article "Digital Trends in the Rail Sector", published in April 2019 and prepared by UNIFE's digitalization platform, UNIFE analyses this relationship [1]. The paper contains full chapters dedicated to big data and cyber security, covering emerging developments such as AI and innovative connectivity technologies. Understanding this relationship can help any company better determine the capability to develop or acquire the data it has to use and to safeguard it.

4.8.1 AN EFFECTIVE CYBER SECURITY STRATEGY FOR THE RAIL SECTOR

As mentioned above, digitalization could undoubtedly help to increase the safety, efficiency, and convenience of rail transport for both passengers and freight. Still, it would also expose rail systems to cyber security risks. While trains use the advantages of digitalization and IoT, cyber-attacks can be expected to be increasingly sophisticated. We should indeed know the versatility and dynamic of cyber threats as the digital world and its applications. There are many different kinds of cyber threats, which can harm rail transport. In some situations, the vital rail infrastructure, for example, signalling devices, may cause physical harm, while in other instances, the rail system may even not be explicitly focused. Identifying and enforcing anticipatory steps is often challenging because of the diverse nature of potential threats. Recognizing and being prepared to deal with cyber-attacks are significant challenges for both the rail industry today and the future. Therefore, the European rail supply sector has led with a robust and secure plan for cyber security. In addition, its members have also drawn up a technical paper detailing crucial cyber security issues in terms of criteria and capabilities, providing comprehensive guidance that directly complements the messages in the vision report. They have also been involved with an engaged "cyber security working group". They have tried to identify, among other things, the following four key elements to design a holistic and sound cyber safety chain within the rail industry:

4.8.1.1 Dedicated Skills

The first step depends on the understanding of cyber risk and its potential rail system impacts. The aim should be to build tailored skills and knowledge relevant to cyber security, as well as raise cyber-consciousness in each business and enterprise about cyber threats. Cyber security related skills should, therefore, be strengthened, in particular concerning two aspects: the detection and response of cyber threats, to minimize the adverse effects of cyber security occurrences and to facilitate a swift recovery of systems and services following such an event.

4.8.1.1.1 Sectorial and Cross-sectorial Cooperation

Cooperation between all the relevant players in the rail sector should be strengthened, in conjunction with the development of critical expertise and the closure of any gaps in skills. The exchanging of knowledge and the sharing of experience with other businesses impacted by this will increase the capacity of railway stakeholders to develop and implement effective measures to protect their network and facilities against cyber threats through existing networks – such as alliances, systems, and government information forums. In particular, at the First Transport Cyber Security Summit, which was hosted both by the European Commission and the European Network and Information Security Agency (ENISA), the importance of collaboration was echoed strongly.

4.8.1.1.2 Security-by-Design

The emphasis on safety aspects during the product design process – giving them the attention required – is also a key element for a successful compliance policy to ensure that relevant laws and regulations can be complied with at an early stage. As a positive result, safety-informed goods and devices will lead to the saving of time and resources, contribute to risk reduction and escape risks of an effective cyber-attack and prevent the hectic and expensive replacement of the components.

4.8.1.1.3 Research and Innovation (R&I)

It is also essential to highlight the fundamental role of R&I in encouraging the digitalization of trains. It also refers to new solutions that can improve the cyber-resilience of rail transport. Safety as a core element is the "Rail 2050 dream" of the European Rail Research Advisory Committee (ERRAC) that stresses the high availability of the rail system and services by a comprehensive ICT infrastructure coupled with reliable corporate continuity processes. In the creation of the Shift2Rail Joint Undertaking (JU) under Horizon 2020, the R&I commitments for the rail sector, in particular, found their field of gravity. On the issue of cyber security, currently, the Shift2Rail JU works on a specific technology demonstrator to achieve an optimal level of protection against a severe threat to telecoms and signalling systems.

The post-2020 period needs a reorientation of rail-based research and development activities, which is why UNIFE strongly urged that the Shift2Rail JU be continued through the Horizon Europe Framework Programme. A "Shift2Rail 2" will enable Europe's rail sector, including in the field of cyber security, to develop several added-value products and services.

4.8.1.1.4 Working Together with the EU Institutions

While the industry is prepared to engage in a successful cyber security strategy, with the European rail supply industry at its forefront, the role and actions of EU institutions are as crucial as ever. The European legislative process has focused on cyber security in recent years, for example, Directive 2016/11 on Network and Information Security (NIS). Also, the EU's cyber security roadmap will include the establishment, through the "Cybersecurity Act" adopted in December 2018, of a network of competence in the areas of cyber security with a European Research and Competence Center and the strengthening of the ENISA. Finally, the General Data Protection Regulation (GDPR) is another landmark piece of EU legislation that concerns cyber security and introduces strict regulations on collecting and managing private information. The GDPR's principal objective is to enhance data protection further and to add new cyber safety solutions. The European rail industry believes that the European Commission will continue to develop a stable, harmonized EU regulatory and administrative structure to identify and address cyber security threats, with the vital support of the European Parliament and the Council. It is necessary to resolve polarization both at an institutional and industrial level in Europe's cyber security environment. In this respect, it will be a positive step in the right direction if ENISA's authority is strengthened to turn it into a real EU cyber security entity. Admittedly, cyber-attack security is a crucial element for safe and reliable railways.

4.9   CYBER SECURITY RISK MANAGEMENT STRATEGIC AND TACTICAL CAPABILITIES

In order to address a broad range of cyber security concerns, cyber security reviews address specific risk management issues, such as operational or managerial monitoring and data acquirement structures. The business supports the aforementioned customers in building ties with management, IT, logistics, and security personnel. It moreover assists customers to classify and alleviate credible hazards to automation coordination like positive train control (PTC), communications-based train control (CBTC), automatic train supervision (ATS), and essential systems. It also helps to identify and mitigate potential risks. Cyber security breakdown agreements include a range of standard amenities to tackle the most difficult questions of cyber security.

Services covered by cyber security analysis:

- The cyber security strategy: development of a methodology for standard-scale risk management that follows best practices from industry and is oriented for customer needs.
- Governance: the concept of acceptable procedures, checks, and balances to ensure cyber security systems monitor danger and threat evaluations.
- Defensive architecture: defensive architecture counters constantly changing challenges.
- Sanitation scheme: project definition and implementation to address security vulnerabilities.
- Wayside, optical camera controls (OCC), communications and signals.
- Cyber security analysis is a forerunner in risk management for rail cyber security.
- Organization knows security-critical systems, forming the preferred cyber security standards of Asia–Pacific Trade Agreement (APTA).
- Cyber security analysis realizes that cyber risk managing includes information protection as well as adequate security alerts. The approaches must deal with technical problems, procedures, and behaviour.

4.10   CYBER-ATTACKS AFFECTING RAILWAYS

Cyber rail threats and other transport infrastructures are no longer a theory. New cyber-attacks have already hit significant trains around the USA, Europe, and Asia. The combination of high vulnerability and severe risk of collapse, economic damage, and even human death makes railways around the world the ideal destination both for economically motivated criminals and for harsh nation-state actors.

(a)   **Cyber-Attack Risk to Railways**

Even when there are several differences, a modern railway uses computer systems to track and control the railway's physical (operative) machinery. These operational technology (OT) systems converge with IT networks where ransomware can quickly be compromised. For most railways, cyber security mainly consists of consumer security products such as necessary firewalls and antivirus software approved by the government. It is analogous to cyber security systems, which are not crucial for national security in most small or medium-sized enterprises. Many industries may find this sort of protection sufficient, but there is no place close enough to shield a highly targeted vital national transport infrastructure from those who wish to harm it. Most rail systems, whether or not openly admitted, have already suffered cyber-infractions. Those who have been able to avoid threats so far realize that their time has elapsed.

(b) **OT Assets that Need to Be Protected Against Cyber-Attack**
There can be a dozen or more OTs, which could lead to significant disruption to railway services if affected by a cyber-attack. This includes the trains, the equipment, and the maintenance of the stations.

(c) **How Can the Railway Network be Better Safeguarded?**
Thales is a market leader in the areas of both rail and cyber protection. The choice of digital technology is ambitious for their customers. Rail companies look to technology for increased capacity, decreased infrastructure costs, and better customer service, regardless of whether it is AI or the IoT, and an increasing number look ahead to the autonomy's revolutionary potential. The clients are ambitious in preferring digital technology. Railway companies aim at increasing capability development, lowering costs for maintenance, and enhancing customer service, regardless of whether AI or the IoT is involved. Thales supports its clients to ensure their enterprise is safeguarded through the protection of critical systems. Benoit Bruyère, Cybersecurity Authority at Thales, explained: "It begins by enforcing adequate security and applying cybersecurity design principles, for example, defense in depth." The identification is then added. Before they spread, we use controlled areas to identify and isolate threats. We are exceptional in our experience in both transportation and cybersecurity. Systems must be patched and secured adequately, and Thales can separate security and safety roles. Patching is essential because policies can keep 20 years or more in service. Via acquisitions – like Guavus, a leader in real-time big data processing – we now benefit from new technologies, such as AI and research. Our approach relies on compliance with commercial safety requirements. This includes conformity to the Industrial Cybersecurity Standard IEC 62443. We establish cybersecurity capabilities in three areas, in addition to providing solutions: monitoring, assessment, and oversight. The fact that our customers face real threats makes these capacities important. With severe economic and reputational consequences, Hackers can disrupt rail services and steal customer data. Security is even at risk for travellers." Bruyère said, "The stakes are high." However, operators can identify threats and react quickly with sophisticated cyber security. They can also avoid assaults. "In addition to reducing risks, improved cybersecurity improves efficiency. This increases asset awareness, allows companies to use open networks, and encourages our consumers to migrate comfortably towards the new railway. Maintenance and patch management services are insecurity," Bruyère said. "To help our customers plan to build a complete security system policy, auditing, and review. We offer cybersecurity surveillance to safeguard the critical

systems of our customers 24 hours a day. However, proper protection gives customers the ability to outsource a reliable partner for security operations."

4.11 RAILWAY CYBER SECURITY: RAILWAY OPERATIONS AND ASSETS SECURITY

To global rail operators, the cyber-attack has been an important issue, and the organization's reliance on connected technologies is increasing. Railway agencies have mechanisms in place that improve their protection, but these reviews still lack a programmatic approach to safeguarding the Indian Civil Service (ICS). A rational and practical counter-media starts on the road to success in cyber security on a rail. In-depth compliance research, comprehensive assessments, and crisis management will improve the strategy and reduce institutional, financial, and technological impacts. After all, protection must be incorporated throughout its life cycle in each component of the solution.

At a moment when rail networks are becoming increasingly automated and rapidly growing, the emphasis is on rail cyber security. Current attacks on the digital system are imperative in rail cyber security technology. Though new digital railways have increased the ability to prevent trains from colliding, have led to improved efficiency, and have made travel quicker and less expensive for customers, the dark side of modern railways can be seen in its cyber security risks.

4.11.1 Cyber Security Railway Future Vulnerabilities

Digital rail protection could be subjected to fatal flaws by studying them more closely [17]. Increased traffic maliciousness could be symptomatic of railway health vulnerabilities. It is prudent since risks are likely to be planned. Because of the transition to open platforms, modular equipment made from industrial off-shelf components, the usage of interacted controller systems, and mechanization processes that can be controlled directly through public and private networks, railway systems are vulnerable to cyber-attacks. The Special Report 800-82, revision 21 of the National Institute of Standard and Technology (NIST) lists the cyber threats posed to ICS [18]:

- Unauthorized modifications to the ordered or warning threshold increasing harm, or disabling or turning off infrastructure.
- Make environmental impacts and endanger human life. Erroneous information leads to managers of the scheme.
- Whether uncovering unauthenticated anomalies or pushing launching operations.
- Improper operations increase or deferred details in distracting rail operations around ICS networks.

With wireless technology and displays in-cab showing speed, signalling systems are increasingly sophisticated on most world railways. The European Railway Traffic Management System (ERTMS) practises the European Train Control System (ETCS), which describes the operation of the rail classification. The control system provides automatic fortification for trains and improves the railway's capacity, security, and operability. However, the digitalization of passengers is opening the door to cyber security discussion – namely, the exposure of the new railways to digital hackers plus cyber-attacks. The threshold aimed at the effect of a cyber-attack is enhanced connectedness.

4.11.2 Cyber Security in the Fight Against Railways

In the face of ever more dynamic cyber operations, the security and safety of travellers and their rail operations are the top priorities. It is important to create, enforce and sustain precise cohesive elucidations, irrepressible networks and value-added services at any time to protect subtle data. Now are several methods of optimally protecting the critical infrastructure of rail operators:

- A monitoring device is a highly productive way in which existing IT systems and devices can, in one screen, identify, imagine, investigate, and respond to threats and liabilities (see Figure 4.11).
- When faced with the possible dangers of cyber security infringements via rail, train operators should disseminate knowledge of the issue and enable rail companies to incorporate mechanisms to track and prevent attacks, and in the case of safety, minimize losses.
- Security infringements are addressed, for example, by Dhavalas Railway Station (DRS), Transportation Security Administration, US Computer Emergency Readiness Team (US-CERT) and their Cyber Emergency Response Team (ICS-CERT) for Industrial Control Systems.
- Supplementary tools such as knowledge on cyber threats and penetration testing to find the "weak spots" where IT structures and procedures intersect.

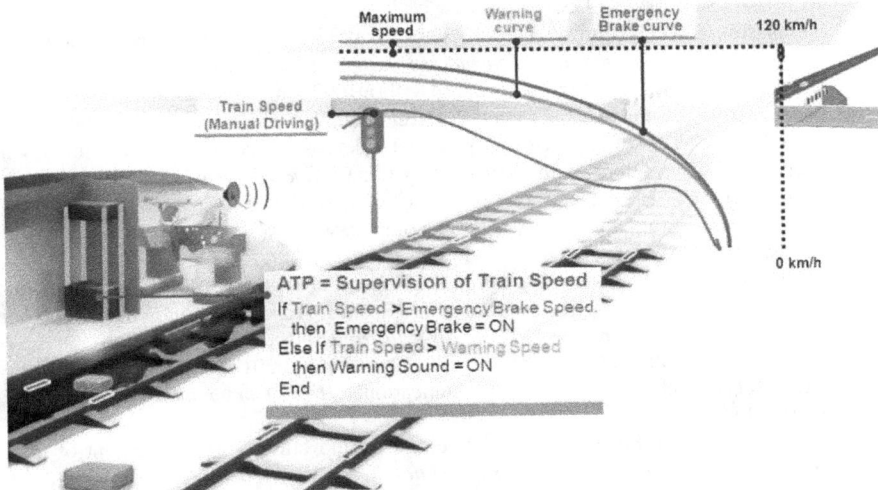

Automatic Train Protection – ATP

FIGURE 4.11 ATP for cyber security.

4.12 CONCLUSION

In all countries around the world, demand for free, reliable, and trustworthy railways continues to be a cause for concern. Operational efficiency and reliability, protection, and intelligence issues, as well as aging railway networks and procedures, are pushing many countries to upgrade their current railway network. The global rail industry faces a problem for the growing demand for freight and passengers because the rail network has not been integrated, and the rail facilities are inefficient. It should encourage rail operators to develop better and more efficient rail systems. IR's passenger reservation system is one of the largest in the world. Every day, nearly one million passengers travel with IR in reserved accommodation. Sixteen million more travel on IR with unreserved fares. In this extensive network, it is a huge task to handle passenger data effectively, which is a crucial issue nowadays. In this article, the authors have discussed various issues surrounding the application of intelligent computation in railway systems.

REFERENCES

[1] Larry T. and Gödel, E. (1999, orig. 1979), Tesler's theorem: AI is whatever hasn't been done yet. In *Bach: An Eternal Golden Braid*, p. 601, Basic Books, Inc.

[2] Lovell, M. (1983), Data mining, *The Review of Economics and Statistics*, 65(1), 1–12

[3] Bezdek, J.C. (1994), Fuzzy models – What are they and why? *IEEE T Fuzzy Sys* 2(1), 1.

[4] Cooper, D.E. and Paul Ezhilchelvan (2004), *Isi Mitrani in Networking 2004*.

[5] Cui, J., J. Zhang, H. Zhong, and Yan Xu (2017), SPACF: A secure privacy-preserving authentication scheme for VANET with cuckoo filter, *Vehicular Technology IEEE Transactions on*, 66(11), 10283–10295.

[6] Ferrari, G., S. Busanelli, N. Lotti, and Y. Kaplan (2011), *Cross-network information dissemination in VANETs, 11th International Conference on ITS Telecommunications*, pp. 351–356.

[7] Huang, X. (2006), "*Smart antennas for intelligent transportation systems*," 2006 6th *International Conference on ITS Telecommunications*, Chengdu, pp. 426–429, doi: 10.1109/ITST.2006.288933.

[8] Stefansson, G. and K. Lumsden (2009), Performance issues of smart transportation management systems, *International Journal of Productivity and Performance Management*, 58(1), 55–70. https://doi.org/10.1108/17410400910921083.

[9] Berenguer, M., A. Palau, A. Fernández, S. Benlloch, V. Aguilera, M. Prieto et al. (2006), Efficacy, predictors of response and potential risks associated with antiviral therapy in liver transplant recipients with recurrent hepatitis C. *Liver Transpl*, 12, 1067–1076.

[10] García, F.P., F. Schmid, and J.C. Conde (2003), A reliability centered approach to remote condition monitoring. A railway points case study *Reliability Engineering & System Safety*, 80(1), 33–40.

[11] Bocciolone, M., A. Cigada, A. Collina, D. Rocchi, and M. Belloli (2002), *Wheel and rail longitudinal wear measurements and diagnostics, Techrail International eWorkshop and Dem on High Technology Systems for Railway Transportation*, Paris.

[12] Ferrus, R., O. Sallent, G. Baldini, and L. Goratti (Oct. 2013), LTE: The technology driver for future public safety communications, *IEEE Communications Magazine*, 51(10), 154–161.

[13] Guan, J.Z. (2014), Scientific and technological innovation and development of intelligent transportation in China. *Inform Technol Stand*, 10, 1–3.

[14] Ferreira, A. and J. Atkinson (October 2005), *Language standard explanatory web dialog, intelligent search officers, IEEE Technology, Special Theme on Smart Search*, pp. 44–52.

[15] Maslekar, N., M. Boussedjra, J. Mouzna, H. Labiod (2011), *VANET based adaptive traffic signal control, IEEE 73rd Vehicular Technology Conference (VTC Spring)*, pp. 1–5.

[16] Mayring, P. (2003), Qualitative content analysis, *Forum: Qual Soc Res*, 1(2), 1–80.

[17] Vinaya Kumara, N. (December 2019), Smart computation based smart card information using rail management system, *Journal of Engineering & Science*, 10(12), 726–732, ISSN:0377–9254.

[18] ICS, The Special Report 800–82, revision 21.

5 Big Data and IoT Forensics

Indu Chawla and Archana Purwar

Jaypee Institute of Information technology, Noida, India

5.1 BACKGROUND AND INTRODUCTION

Big data is a growing field where advancements in innovation take into consideration better approaches to manage colossal amounts of data continuously being generated by an immense assortment of sources like Internet of Things (IoT) [1]. This term has been instituted as different objects attached to the network generate and trade an incredible quantity of different types of data at speed [4, 5]. Big data alludes to huge data sets of organized and unstructured computerized data gathered from various sources [6]. Sensors and radio frequency identification (RFID) systems, cameras, and various other devices are considered as the most significant hotspots for spawning big data [7]. The IoT is additionally responsible for producing a large volume of data [9]. It is said to consist of everything that can move or get data and be connected with one another. The IoT could be considered as extended and inescapable in every domain. This vast communication system is fit for assembling, preparing, and transmitting data to investigate each network component. The IoT foretells that the world would be constrained by different devices (things) that are associated with one another through a solitary infrastructure [8]. As such, the IoT is a giant system in which an immense sensor, RFID, mass of things, and smart gadgets are capable of interacting with one another [10].

Developing big data produced from the IoT represents an assortment of issues in the territory of digital forensics that is a significant instrument to recognize the crimes that are inherently using computers [2]. Digital forensics is universally used inside law authorization to research electronic media and progressively inside associations as a feature of their incident response procedures [3, 11]. Digital forensics is getting all the more testing for the reason that the massive increase in registered smart devices and PCs has enabled its expansion globally. This increase is presenting new difficulties when we process the digital data in a distributed and parallel fashion for a better response. The expanding usage of cloud facilities benefits companies in their day-by-day tasks and the increased improvement of smart gadget use implies that examination of digital evidence including such frameworks would include progressively complex digital proof procurement and investigation [12]. In the presence of virtual environments offered by cloud technology, digital criminological examinations could be very problematic because of the dynamic idea of data with the approach of IoT gadgets.

The IoT is a developing idea, with no standard design embraced consistently, although numerous data analysts and researchers have suggested different models to delineate conceivable methods for realizing it. Atzori et al. [13] realized the architecture of the IoT in three paradigms with three different perspectives: the Internet, things, and semantics. Additionally, the IEEE P2413 project for the IoT likewise suggests the design is a three-level framework comprising applications as the first level, networking and data communications as the second level, and sensing as the third level [14]. Gubbi et al. [15] characterized the IoT as an interconnection of detecting and actuating gadgets that is able to impart data across platforms using a built-in framework. This framework constructs a typical working picture to enable pioneering applications. Also, Khan et al. [16] proposed a wide-ranging model of the IoT comprising five tiers. These are perception, network, middleware, application, and business. The bottom tier in this architecture is taken as perception, while application and business are considered as the outermost tiers. From the view of IoT forensic investigators, Parag H. Rughani [17] characterized a general architecture having three layers as "end points", "network", and the cloud/server layer. Most of the IoT realizations are structured around their framework in one way or another. Digital evidence, which might be anything varying from sensors to microcontrollers, smartphones, RFID tags or laptops, are present in the bottom layer (end points). This layer is the last point of the system. The intermediate layer (network) is essential because it permits end points to interact and share data with each and every one. It also helps to preserve data in cloud storage. Lastly, the upper layer (cloud/server) is not mandatory in some cases, yet most IoT implementations may use this layer for preservation of data required in future.

5.2 TYPES OF IoT FORENSICS

In the time of IoT, various digital gadgets like automobiles, smoke alarms, watches, glasses, webcams, and so on are coming together using the Internet connection. The quantity of devices that have the capacity for remote access to screen and gather information is ceaselessly expanding. This advancement makes the person's life increasingly agreeable and advantageous, but it additionally introduces openings and dangers for the digital forensics analyst from a crime viewpoint [18]. Computer analysts in forensics manage the detection, sorting, examination, and production of evidence in digital form. This digital evidence is collected from different sorts of digital or electronic storage space involved in a different cybercrime or data security event [19]. Normally, storage media range from a server to a mobile phone. The origination of IoT devices and the expanded number of crimes using computers/servers on IoT gadgets/applications has demanded the assortment and examination of computerized evidence coming from various sorts of IoT gadgets. IoT forensics is termed a subcategory of computerized forensics that demands a versatile methodology in which proof might be gathered from an assortment of IoT sources. These sources have sensors, actuators, specialized gadgets, distributed storage, and even ISP logs that produce data independently. This data could also be the result of some human activities such as movement of people, entry of a person, and others. This continuous and active data is superb for digital analysis as they are able to catch every activity, which is of use in examinations. IoT devices are priceless sources of digital proof, providing data along with device information to forensics specialists. Digital evidence can be gathered from sensors, smart devices, and internal networks (for example, a firewall or a switch/router), as well as outgoing networks (for example, the cloud or an application). In light of these zones, IoT forensics can be arranged into three sorts [20].

5.2.1 CLOUD FORENSICS

Cloud forensics is a subcategory of computerized forensics representing one way to deal with crimes in the cloud environment. Cloud forensics can be found in two different ways: client forensics [27] and cloud server forensics. Client forensics manages proof data, for example, session data, registry data, cookies, history logs, access logs, temporary data, and chat logs. It can be easily captured from the web browser [28]. Identifying and collecting evidence from the client side plays a vital role in investigation [29]. This data is significant and must be gathered as soon as conceivable in its sterile state for presenting as proof to solve criminal cases. There is a high probability that the data could be deleted either deliberately by a user, or incidentally by crackers.

Cloud server forensics incorporates a variety of logs such as access data, application logs, database logs, user authentication logs, systems logs, and so forth. As the resources in the cloud are difficult to access and locate, it makes it a lot harder to perform the identification, division, and assortment of computerized evidence in cloud forensics. IoT forensics in the presence of cloud technology encounters various challenges due to three distinctive service providers. These three major providers are Software-as-a-Service (SaaS), Infrastructure-as-a-Service (IaaS), and Platform-as-a-Service (PaaS). These providers encourage sharing of their assets in the cloud so that IoT gadgets may cross the Internet (immediate or aberrant interface) through various applications. As most of the data gets stored in the cloud, it has recently become the main focus for intruders. In customary computerized forensics, the analyst is able to cling to the digital gadgets and afterward relate the examination procedure to find the proof. However, cloud forensics [21, 23] allows the digital evidence to be stored in different locations across the world, which is escalating numerous difficulties regarding procurement of digital evidence from the cloud. Likewise, in the cloud, forensic analysts possess restricted command and retrieval rights to hold onto the digital gadgets because of various service models, for example, PaaS and SaaS. These cloud providers do not give power over equipment. Additionally, in the event of a cyber-crime, a court request given in a place where a data community dwells will probably not be relevant to the purview for an alternate host in another nation. Hence, it is hard for specialists to choose appropriate proof [22]. Additionally, data could be stored in an alternate area in the cloud, so no proof could be seized. Moreover, cloud administrations utilize virtual machines as servers, so unstable data like cookies, web logs, and temporary files could be eradicated in the event that they are not coordinated with storage devices such as Gmail.

5.2.2 NETWORK FORENSICS

Numerous devices are connected using different kinds of network, namely local area networks (limited range), metropolitan networks (restricted to metro cities), wide area networks (across countries), wireless networks, and others. These devices keep on sending and receiving messages or data over the network. Various logs such as firewall logs, intrusion detection system logs, and others are created during transfer of data across the network. These logs might be collected as probable evidence for the processing of criminal cases [24]. This kind of forensics is able to manage the investigation of system traffic and logs to track the different events involved in criminal cases, and this is theoretically also feasible when cloud as well as IoT technology is being used. The diverse layers in the Transmission Control Protocol/Internet Protocol (TCP/IP) model are capable of offering the required information on communications between various virtual machines within the cloud [27]. Although the network traces or communication logs are crucial for forensics purposes, these important logs produced by customers' events or applications are not provided by cloud service providers (CSPs). For instance, on the off-chance that somebody utilized an IaaS instance to disseminate malware, network traces as well as routing information are critical pieces of legal data; however, they are hard to acquire. This turns out to be all the more trying in different cloud systems like PaaS and SaaS. Hence, the collection of crucial data relies intensely upon the support provided by the CSP to the forensic analyst.

5.2.3 DEVICE-LEVEL FORENSICS

Device-level forensics incorporates all potential evidence in digital form which can be gathered from different sorts of gadgets, for example, fax machines, iPods, personal digital assistants, laptops printers, digital cameras, or pen drives. These provide various sorts of data like images, text, sound, and video [25, 26]. Device-level forensics may include audio and video obtained from Amazon Echo and video surveillance system footage, respectively. These can be incredible pieces of digital evidence in investigations. Mobiles telephones are additionally significant gadgets that have become versatile data bearers, and they monitor every one of your moves, email, making and putting away reports, recognizing areas with GPS benefits, and overseeing business errands. Data gained from telephones can be turned into a precious source of proof for examinations associated with criminal, common, and even prominent cases. It is uncommon to lead an investigation that does exclude a telephone. Cell phone calls logs and GPS data are utilized as digital proof.

Smartwatches and embedded personal devices can likewise be an important source of data for the examination of criminal cases, for these are capable of storing call records, contact details, texts, images, video, and so on. Moreover, these devices do not require any access to a connected smartphone [30]. At present, commercial forensic tools can carry out the securing and scientific examination of an extremely modest number of smartwatches. No particular tools in the market target the procurement of the data contained inside these Real-Time Operating System (RTOS) smartwatches. However, it is essential that the forensics investigators should recognize the smartwatch to identify the model of Mediatek (MTK) chip used in a particular smartwatch, on the grounds that based on the model, it is possible to identify which operating system is used. By knowing the specific operating system, such as Android, Android Wear, watchOS, and so on, different forensic tools could be used to gather and study data. In the event that smartwatches utilize a Nucleus RTOS, the method suggested by Gregorio et al. [30] can be used received for various procedures involving collection, investigation, and generation of results.

5.3 SOURCES AND NATURE OF DATA

The IoT speaks to the consistent converging of the real as well as computerized world, with new gadgets that save and process various data. It creates a lot of problems for forensic analysts. It results in IoT forensics, which is termed a subfield of computerized forensics. IoT forensics emerged as an ongoing necessity to give an all-around organized and logically effective method of criminology examination to take care of digital crime. Cybercrime is a socially inspired advanced wrongdoing arising from misuse of the network. The multifaceted cybercrime space has broadened from straightforward identity dangers to geopolitical wrongdoing over a few decades. Reports from a cybercrime branch of Europol discusses digital detection of crime, which involves gathering and analysing data involved in crimes, aside from proof for examination. It is likewise an information hotspot for setting up preventative procedures and understanding the thinking processes of the criminals [31].

Typically, cyber forensics processes can be visualized in three stages, namely collection, investigation, and dispensing of evidence. Cyber forensics analysts are facing a major challenge due to the advent of IoT technology and the Internet of Anything (IoA) [32]. The IoA includes cyber–physical systems, smart grids, drones, intelligent buildings, smart swarms, and autonomous and digital organic frameworks. The IoA is strengthened with different technologies such as big data, mobile computing, and pervasive computing. In fact, IoT technology is heading towards insightful intricacy but this is more difficult for the IoA. The IoA takes the whole thing "on the web" in a connectionless fashion that creates a blast of attached connected smart gadgets ranging from cars, refrigerators, drones, and vehicles to smart grids, swarms, buildings, and cities. With adaptable IoT gadgets, there is no particular technique for IoT crime scene investigation that can be comprehensively utilized. It requires distinguishing strategies for performing IoT-based computerized criminological investigation. With a normal quantity of 50 billion objects by 2020, attention should quickly be paid to communicating, amassing, accessing, and treating the gigantic volume of information made by these devices [33]. The different sources of data in a smart city/smart world in cyber forensics investigation are smart cars, smart homes, sensors, unmanned aerial vehicles, smart buildings, and smart grids [34]. Broadly, data gathered from a variety of sources is categorized into two types.

5.3.1 BIG DATA

Big data is propounded as a pool of potential knowledge, which is used by organizations for finding implicit information and structured analysis for better outcomes. When different industries, users, and governments wish to exploit information with online as well as offline data, they inexorably look forward to solutions provided by big data technology, one of the most valuable assets in today's world. The amount of information generated every year is increasing faster than any time in recent memory. By 2020, each human on the planet will make 1.7 megabytes of data every second. The International Data Corporation (IDC) forecasts that the global datasphere would increase from 33 zettabytes in 2018 and to 175 zettabytes by 2025 [35].

Big data is typically enclosed by three different kinds of data known as unstructured, semi-structured, and well-structured data. Unstructured data includes tweets, emails, text, audio, video, images, or other social media posts that are text-intensive and cannot be simply interpreted by conventional models. Semi-structured data appears to be like structured data but it contains markup tags or elements to distinguish different types of elements. It also creates hierarchies of a variety of records and fields present in the data. Such data includes web data in the form of cookies, Extensible Markup Language (XML) data or JavaScipt Object Notation data. Structured data refers to the data stored in relational databases, which has a predefined format. Such data includes tabular data in the form of columns/rows, comma-separated values in records, and others. Typically, the six Vs given below describe big data.

- Volume: It indicates the capacious amount of data spawned by distinct objects connected through communication media.
- Variety: It shows several types of data having different modalities like images, audio, video, blogs, and others.
- Velocity: It propounds the pace at which data is being produced by different devices and with the advent of advanced communication technologies.
- Veracity: This term refers to any kind of abnormality, uncertainty, noise, or bias in the data. To extract meaningful and reliable patterns from the data, its source, type, and processing should be reliable and done carefully.
- Variability: This term shows that data is continuously changing with rapid speed due to different sources such as sensors, RFID, social networks, and others. Hence, data solutions must be capable enough to store, analyse, and extract the relevant data. Data pre-processing must ensure that only valuable data should be used to increase the efficiency of the solutions/applications without any information loss.
- Value: This refers to only important information extracted from the big data, which is useful for different organizations, societies, consumers, and entrepreneurs, and provides many benefits.

Big data needs to be converted into useful information to assist industrialists, e-learners, scientists, astronomers, and others. Typically, it is spawned using three major components, namely day-to-day data, data generated by IoT devices, and data resulting from online social media. This term has brought a drastic change in academia as well as industries.

5.3.2 IoT Forensics Data

Due to advancements in technology and network speed, the Internet is continuously changing with certain new sorts of software program and hardware. Most people all over the world are using the Internet in some way or another [35]. The exchange of information may be both human–device communication and human–human communication due to IoT technology. This advancement contributes a major part of the big data generated by IoT devices. An IoT gadget keeps a sensor connected to it and is able to send data from one thing to another thing or to humans using the Internet. It includes PCs, software, wireless sensors, actuators, and others. These are connected to a specific object that works using a wireless/wired network connection, allowing transmission of the data between devices or persons automatically, without the individual's intervention [36].

These IoT gadgets produce traces that can be valuable for analytical and criminological purposes when an offence has occurred. For example, locating IoT gadgets around people goes amiss on occasions, including homicide, as computerized proof can be gathered after the wrongdoing has occurred. In a perfect world, obviously, the framework would caution the police that a wrongdoing will happen by using the prediction models, to prevent the wrongdoing from occurring. IoT crime scene investigation information could be gathered using pre-set sensors in homes and structures, sensors incorporated with movable vehicles and gadgets, distributed storage, and web access providers. Generally, during forensics examination, IoT gadgets could be dissected in following stages [18]:

Preliminary Analysis
At the point when an analyst realizes that a fresh device is attached to the network, the initial stage is to overview the ongoing work on the sources, encompassing scholarly work, safety features such as data vulnerability, and social network sources. Preliminary analysis provides data about strengths and weaknesses of the attached gadget, as well as potential approaches to reach the gadget.

Network Analysis
IoT devices attached to the network need to be tested extensively within a controlled environment with settings to facilitate forensics investigations. This testing environment needs a network configuration which provides an inactive collection of traffic to and from connected devices. Moreover, the network configuration must facilitate man-in-the-middle (MITM) attacks for testing whether some malicious activity is going on or not. The target of examining system traffic in the test condition is to consider the interchanges with the IoT gadget involving communication between different gadgets or frameworks, correspondence conventions used, and data transmission, either in plaintext or encoded, which is powerless against MITM assaults. Inspecting the diverse

routes into the gadget, the listening ports, and the dynamic administration will give bits of information about potential approaches to get data from the gadgets, or the accessibility of distant access administrations, for example, SSH or Telnet.

Analysis of Smartphone Applications

IoT gadgets, particularly intended for smart homes, give systems to the customer to screen and control the gadget by means of the Internet or nearby network. These customer communications normally include a cell phone application that can save data about the gadget, its set-up, and past records of different activities. Investigation of the IoT and related cell phone software programs needs examination by human intervention, as mobile-based forensic tools do not, largely, incorporate parsers for these specific applications. Top to bottom, manual examination of the cell phone application, including figuring out, can reveal extra data, which can be related to occasions traced by the IoT gadget and directions communicated by the client inside the test condition. The consequences of this top to bottom examination can be arranged by composing custom modules inside open source digital forensics devices to consequently process the leads. These modules can be accessed again when the gadget is encountered in prospective examinations.

Analysis of Vulnerabilities of Devices

It is critical to inspect susceptibility of the IoT gadget to see how a gadget could be undermined and abused by a harmful person to execute a crime, in order to find potential approaches to access the gadget so as to get data. Examination of IoT gadgets' vulnerabilities consolidates the entirety of the data from the beginning, and investigates the majority of widely recognized vulnerabilities for the sources found [37].

Physical Analysis

When researching criminal cases, it is generally unrealistic to retroactively arrange the logs generated during network traffic identified with a relevant IoT device. In such cases, the data stored on the IoT gadget itself is of principal significance. Along these lines, the last advance in the examination technique for IoT gadgets is to carry out a physical investigation of the equipment. Contingent upon the IoT gadget, it could be conceivable to pass through a sequential connection (UART), and it could also be important to continue to chip-off systems to get to the gadget memory.

Cloud

An added advance when managing IoT gadgets in forensic analysis is to get related data from related cloud service organizations. Data from IoT gadgets is regularly stored in the cloud for simple recovery by related cell phone applications. This data may be accessible by the cloud specialist organization or by utilizing cloud identifiers recovered from the client telephone. Lawful approval is regularly an essential for getting such data in the cloud [38].

5.4   ROLE OF BIG DATA IN IoT FORENSICS

The arrival and expansion of intelligent and smart devices and the data security cases that arise from these devices have required their criminological categorization and investigation through the act of IoT forensics. In IoT forensics, devices continuously generate abundant amounts of data that cannot be stored in the form of tables and simply queried by structured query language (SQL). It necessitates the use of big data technologies facilitated with IoT analytics to make the intelligent decisions in forensics investigations. A number of the big data frameworks along with big data analytics are listed below.

5.4.1   BIG DATA TECHNOLOGIES

Big data technology can be characterized as a software utility that is intended to explore, handle, and extract the information using tremendously complicated and huge data sets coming from IoT devices, as in the case of network forensics. The existing techniques for IoT forensic analysis show in-built issues when it handles voluminous, diverse, and streaming data coming at high speed. Hence, digital forensic investigations have become a prolonged and resource-consuming process. The following technologies can assist such analysts to reduce time spent and deliver results much faster.

5.4.1.1   Hadoop

Hadoop [54] framework has been developed to handle the various issues with big data. It is an open source system that has potential to store heterogeneous and huge amounts of data and provides a solution for distributed and parallel processing to make the computation faster at low cost. This framework comprises two major components, namely the Hadoop distributed file system to accumulate data and the map reduce based programming model for processing the data. Further, the resource management in a cluster is typically handled by its resource manager YARN.

However, these are not sufficient to tackle big data problems. Along with these, the Hadoop framework encompasses various other elements such as Hive, Pig, Mahout, Sqoop, Flume, Zookeeper, Oozie, and others to provide a complete solution to big data problems.

5.4.1.2   Spark

Spark [55] is not the same as the Hadoop framework as it can be integrated with different file systems such as HDFS, Mongo DB, Amazon's S3, and others. It is an open source cluster computing system, which is able to run data on a disk as well as in memory. Spark is built to execute onto HDFS and can use YARN. It is developed to do a lot of analytics by employing SQL, streaming, graph processing, R programming, and machine learning algorithms. The fast processing power of the Apache Spark is because of its in-memory cluster feature. In fact, it can run an application a hundred times quicker in memory and ten times quicker when executed on disk. Spark is powerful and very popular as it has a built-in application programming interface (API) in diverse languages such as Java, Python, and Scala, and it provides around 80 high-level operators for users' queries. It is a single integrated software that provides enormous functions for advanced data analytics; it has the ability to read data from any existing Hadoop data file or HBase or Cassandra, and many other data sources.

5.4.1.3 Kafka

Kafka [55] is an open source stream processing platform that can quickly process vast amounts of dynamic data generated from IoT devices. This platform is an intermediary between the source and destination to keep the data for a specific time duration and then process it. It has a distributed messaging system that enables communication of messages from source to destination. Kafka is appropriate in cases of offline as well as online messaging. The messages in the Kafka platform are preserved on the disk and duplicated within the cluster so that data cannot be lost. Kafka is designed on top of the ZooKeeper. It can be well integrated with Spark and Storm to facilitate the real-time streaming data analysis required for IoT forensics investigation.

5.4.2 Big Data Analytics

The data is gathered from various items. The capacity to dissect a tremendous amount of IoT data helps specialists in managing a lot of data that could impact the examination. With this stated, in any case, the more complex technique of handling IoT big data implies that it is hard to easily examine the data that is accessible for the examination. There is surely a critical need to examine new methodologies on how precisely to handle the enormous quantity of data. This gigantic amount of data could be normally dealt with by big data analytics that processes a lot of for the most part human-created data to help longer cases. It includes use cases such as capacity planning, floor planning, target planning, air-station planning, and revenue protection and others. Meanwhile, IoT analytics aggregates and compresses huge amounts of machine-generated data that has low latency and low duration. This data originates from a vast variety of sensor devices to assist various time critical use cases like traffic light operation, portfolio optimization, fraud detection, real-time ad bidding and security violation recognition. Big data corporations are answerable for dealing with enormous amounts of data originating from different channels by giving answers to the following issues looked at by IoT investigations:

- Providing a flexible and scalable method to handle fast and large amounts of streamed data by moving to a PaaS model.
- Providing security due to a pool of varied types by creating a checkpoint to check the connected equipment joined to the network.
- Developing protocols like Mosquitto to offer a controlled mechanism and leak-proof technologies to handle huge amounts of data sent over different channels such as Wi-Fi, GPS, and Bluetooth.

The IoT and big data are interrelated with one another. The IoT will produce colossal quantities of data that must be explored so that IoT systems work without any problem. The devices linked in the network may produce excessive data, leading big data analytics to investigate the relationship between various features of data, find useful results, and track the trends in the data. These trends and results may influence emphatically or contrarily the forensics analyst for criminal cases that may require the forensics examinations. As a result, new algorithms and proficient techniques for data mining, like big data machine learning, are required. Comprehensively, it is assembled into three subcategories: supervised learning, unsupervised learning, and reinforcement learning [39]. Big data machine learning incorporates the following kinds of learning.

5.4.2.1 Data Stream Learning

The forensic investigations consist of big data coming from different IoT devices. Data mining techniques are critical to find potential and interesting examples and to extricate uncovered value from enormous and streaming data sets. Nonetheless, conventional data mining approaches such as descriptive and predictive analytics, suffer from the problems of scalability, efficiency, and precision when utilized for huge and dynamic data sets. On account of the dimensionality, pace, and fluctuation of streams, it is not achievable to store them forever and then to analyse them. Along these lines, researchers aspire to discover better approaches to optimize existing solutions to produce accurate results by processing data samples in a timely manner with limited memory of IoT devices. Besides the time-variant nature of stream data, it also deals with the concept drift problem. A change in the distribution of data over time is known as concept drift. Various experiments on data streams have shown degraded performance of classification models. Subsequently, a few data mining approaches like classification and clustering were modified to incorporate drift identification methods in a dynamic environment. Trials on data streams showed that a change in the fundamental idea influences the presentation of classifier model. Along these lines, improved algorithms are designed to identify and adjust to the idea drifts [40, 41].

5.4.2.2 Deep Learning

Nowadays, data to be investigated in IoT forensics is of multiple modalities. It can be video, audio, images, or text. Hence, deep learning techniques are an extremely hot and active area of research in various fields of artificial intelligence such as machine learning and pattern recognition. These approaches have a significant role in forensics applications for discriminative tasks, namely classification and prediction. They include face recognition, fingerprint recognition, handwriting recognition, cloth recognition, credit fraud detection, anomaly detection, and others. These methods are all the more dominant to determine data diagnostic and learning issues found in immense data sets. Actually, these approaches are able to extract unknown or hidden data patterns from labelled as well as unlabelled data. Further, as deep learning algorithms are dependent on hierarchical learning, it can be employed to make semantic indexing, data tagging, and information retrieval from large data volumes simpler.

Deep learning provides typically two types of neural network: discriminative and generative models [46]. Discriminative models require the labelled data. They include convolutional and recurrent neural networks. Generative models do not require labelled data. Typically, they include recurrent neural networks, deep belief networks, deep auto encoders, deep Boltzmann machines, and others. Recently, researchers have designed various deep learning solutions for application in the area of network forensics [42–45].

However, vast amounts of data in IoT forensics impose a major issue for this set of learning algorithms. Hence, it necessitates tackling a vast number of objects (inputs) and multiple types of class labels (outputs), as well as high dimensional data generated by different IoT sources. The training phase is a very time-consuming task for deep learning algorithms due to the huge amount of data and hardware requirements. It results in the designing of efficient and scalable parallel methods that can optimize the training phase of deep learning methods, which can reduce model complexity and time complexity. Another major challenge is the presence of incomplete data and data having noisy labels. Lastly, the high velocity of data that has to be processed in real-time poses another issue, as data is often dynamic and changing distribution over time.

5.4.2.3   Incremental and Ensemble Learning

As the data coming in IoT is very fast and changing over the time, incremental and ensemble learning methods can be applied to find the useful patterns present in the data. This big stream data collected from IoT devices needs some dynamic learning as it faces the problem of concept drift [47]. Concept drift arises because classification boundaries and cluster centres keep on changing due to high stream data. This problem can be overcome by using incremental as well as ensemble learning techniques. These approaches have the ability to deal with various issues such as limited storage, data availability, and others with IoT technology. The classification or forecasting is done faster when it receives new data using incremental learning. This learning is actually inherent in many traditional machine learning algorithms. These include neural networks, decisions tree-based classifiers, Gaussian, radial basis function neural networks, and the incremental support vector machines [48]. Ensemble algorithms are more apt to incorporate the concept drift problem due to its flexible nature. In fact, nearly every classification algorithm can be used in ensemble learning, while every classification algorithm cannot be used in incremental learning [47]. As a result, incremental algorithms are suggested to be used if the concept drift is smooth or if it is not present. In addition to this, incremental learning is more appropriate when we have a less complex data stream that requires intensive processing in real-time. On the other hand, ensemble approaches can be used to get accurate results when there is huge concept drift or abrupt concept drift. Moreover, these approaches should be used when a complex or unfamiliar distribution of data streams is present.

5.4.2.4   Granular Computing-Based Machine Learning

Granular computing-based machine learning [49] has of late become increasingly well known for its utilization in different big data areas, for example, digital security [53]. It shows numerous favourable results in machine learning, data analysis, recognizing potential patterns, and uncertain reasoning for the enormous amount of data involved in forensics investigations. Granular computing [52] is based on strong foundations or granules, for example, clusters, groups, classes, subsets, or intervals.

In this manner, it might be utilized to develop/design an effective computational model for complex and big data applications, for example, data mining, analysis for digital proof, remote sensing, and enormous databases of interactive media and biometrics. This computing gives useful and efficient tools to analyse data at multiple granularities from multiple views in IoT forensics. In addition, various strategies in granular computing have the ability to work efficiently for crisp intelligent systems as well as fuzzy intelligent systems in a dynamic environment. Granular computing empowers analysts to handle intricate issues (for example, evolution of new attributes/objects over the time) concerning streaming data by giving cost-effective solutions. Various concepts of uncertain decision theory, such as random sets, rough sets, and fuzzy sets, can be used in granular computing. Also, these concepts are applicable in different phases of the big data value chain. At the initial stage, it handles uncertainties present in data. After that, it annotates data and lastly provides granular representation of data, which is fed to machine learning algorithms for decision-making purposes. Consequently, these techniques can tackle big data challenges by employing pre-processing approaches or by rebuilding the problem at some granular level. Recently, various studies [50, 51] have developed a granular-based classifier.

5.5 IoT FORENSICS INVESTIGATION FRAMEWORK

The investigations facilitated by the data collected from IoT devices need to undergo various stages. So, a framework is required, which will ease the task and guide the workflow required in serving data in an investigation [54–57].

5.5.1 STEPS FOR IoT FORENSICS INVESTIGATION

5.5.1.1 Evidence Collection

The first step in IoT forensics would be to collect the data required in an investigation. The collection phase in the context of IoT forensics represents sources of evidence. There are many IoT devices present in a smart environment. First of all, these sources need to be identified. Although the whole process of investigation starts after the crime is committed, the IoT devices present in the smart environment should be forensics ready, meaning the devices should be planned and prepared in such a way that they can be helpful in further investigation. The collection phase has a major responsibility as any presence of errors in the evidence collection phase may lead to the wrong type of investigation and false results.

5.5.1.2 Examination

A digital investigation process is commenced in response to any incident occurring within an IoT-based environment. The examination starts with incidence detection, and collecting resources and related data.

5.5.1.3 Analysis

The data collected thus far is analysed using the expertise of a forensics investigator. The investigator may select more data if required. The relevant data is tagged. It is recommended to create a centralized format to analyse the occurrence of events during a time framework. A timeline analysis is done to study the data and figure out sources which can contribute as evidence.

5.5.1.4 Reporting

The output of the analysis phase presents a set of collective evidence that shows the occurrence of some incident along with interpretation, explanation, and reporting. The output can be shown in the form of reports or in graphics format. Various artefacts can be shown graphically by accessing different attributes present in the data set. The appropriate attributes need to be selected depending upon the investigator's requirement. Along with data sharing, data security is also important. Sensitive data might be present; in that case, irrespective of sharing original data, device metadata can be shared.

All four steps are the backbones of digital forensics investigation. There is an urgent need to address evidence collection, examination, analysis, and reporting measures for IoT forensics in different applications. While dealing with IoT forensics, the forensic soundness must be ensured.

5.5.2 FORENSIC SOUNDNESS

Forensic soundness refers to the fact that the evidence collected for the investigation process must follow certain criteria. The evidence presented must be reliable and complete when presented before the court of law. Forensically sound evidence will ensure the reliability of the IoT device [58, 59].

5.5.2.1 Meaning

This criterion states that the meaning of the evidence collected from the device should be intact. The data in the device must be kept in the state in which it was found. The data should not be changed while in the process of collecting the forensic evidence.

5.5.2.2 Errors

There may be some inherent software and hardware errors. They must be carefully removed, keeping in mind that they do not impact the quality and validity of evidence. There must be explanations for the errors that occur and the need to remove those errors.

5.5.2.3 Transparency and Trustworthiness

The digital evidence collected from smart devices should be reliable and reproducible. Under the same circumstances, the device should be able to produce the same observations. Moreover, the transparency and trustworthiness of the evidence can be shown by mentioning the steps taken to collect the evidence and disclosing the environment in which the analysis is carried out, along with the hardware and software used.

5.5.2.4 Experience

The digital forensics evidence collection process should be carried out only by the forensics expert. A forensics expert is familiar with the properties to be maintained and knows the criticality of information for the process. The evidence collected should be represented with the fact that the resulting evidence data has not been impacted and is forensically sound.

5.6   CHALLENGES IN IoT FORENSICS

The presence and availability of large-scale data generated from IoT devices has created an impact in the investigation field. However, many challenges exist while investigating using digital forensics, as the currently available tools and methodology of digital forensics do not fit with the IoT devices infrastructure [60–62].

The presence of a large number of IoT devices around us makes it possible to produce massive amounts of data in the form of possible evidence. Simultaneously, it will present a challenge for data management investigators to deal with the vast amount of possible evidence coming from different sources. Moreover, with the possibility that the amount of evidence is so big, it is hard to identify the required information that can be presented as evidence in digital crime [63]. This section depicts the current challenges required to be addressed with respect to IoT forensics.

5.6.1   Variety of Data

Along with the massive amount of data generated from a lot of diverse IoT systems, the data received is also of heterogeneous format. Moreover, IoT devices are connected to various nodes working on different technology and following different communication standards. It is very hard to gather data from these sources and further use them for IoT forensics.

Another concern related to the massive data generated through IoT devices is the storage limitation in IoT devices. Cloud storage is used to store the data generated by IoT devices. For IoT forensics, the data needs to be collected from the clouds. In general, the data extraction process from cloud is challenging as the data is distributed among numerous data centres and it is not known where the data is located.

5.6.2   Security

The data available from IoT devices can also be tampered with, hence the reliability of potential evidence is of great concern. The number of IoT devices increases human comfort but it also raises questions related to security of personal data.

The security is related to authenticity, authorization, confidentiality, integration, and availability. In the context of IoT forensics, the security requirements may be different from the traditional requirements. The security threats can be from the malicious user, who takes control of the device, steals the information, introduces some malicious code, or changes the content present. It can also be from the manufacturer of the IoT device, who knows the device and can access the user information and pass it on further [64].

5.6.3   Privacy

The collection of data from heterogeneous sources of IoT devices in IoT forensics plays a big role but along with this the user is an important stakeholder or participant in this data. Collection of possible evidence for forensics investigation should preserve the privacy of individual. The proper utilization of an individual's information is a privacy concern.

Privacy preservation implies protection of private data during the data collection phase as well as during the analysis and results preparation phase. Protecting data at the point of evidence collection is called input privacy. The preservation of personal information during investigation and fact presentation is called output privacy.

In data mining, privacy is preserved by the use of privacy preserving data mining techniques. For input privacy, methods like k-anonymity, l diversity, t-closeness, and differential privacy are used [64]. Similarly, to preserve output privacy, methods like association rule hiding and query auditing are used. Cryptography-based techniques are also very popular to preserve privacy. According to Nieto et al. [65], in IoT forensics, privacy should be maintained throughout the life cycle of the forensic process. This includes the data collection, preparation, and examination as well as analysis. Finally, the digital evidence should be removed after the investigation process is over and the case is solved.

5.6.4 DATA ORGANIZATION

The purpose of building IoT devices was to assist users and to provide connectivity among various devices and applications. However, the quick expansion of IoT devices introduces many fresh challenges in terms of IoT forensics. The large number of IoT devices produces huge amounts of data or information which may act as possible evidence. In IoT forensics, logs from different devices are used for investigation purposes. Application logs, process logs, and network logs are analysed to figure out the presence of any malicious activity. The IoT devices created by different vendors use wide varieties of data structures, which make the analysis and examination task tedious. There is a strong need to address this problem. A standard log format will make the task less troublesome. Vendors should follow a standard protocol to make the IoT forensics less complicated.

5.7 CASE STUDIES USING IoT FORENSICS

5.7.1 SMART HEALTH MONITORING SYSTEM

IoT devices are also used in healthcare applications where they store data regarding patients' health conditions. Due to privacy and security concerns, this information is very sensitive. Any malicious activity affecting this data may be life-threatening for the patient. Moreover, there are many IoT devices in a hospital network. Data collection from all the devices for forensics investigation in a short time period is a critical task.

Hossain et al. presented a case study which shows how an attacker can steal patient information and perform a malicious activity [66]. The patient's mobile phone is connected to a blood sugar monitoring device and other smart devices present in the person's home. Simultaneously, a person who is working in the hospital and the phone are also connected to many smart healthcare devices present in the hospital. The attacker introduces malware and infects the blood sugar monitoring device as well as healthcare devices in the hospital.

5.7.2 AMAZON ECHO AS A USE CASE

Some IoT devices are found in civilian settings like the home environment. Amazon Echo and smart fitness bands worn by victims are examples of IoT devices in a home environment [70]. Amazon Echo is widely used as a voice guarded hub consisting of various smart sensors which respond with the wake word – Alexa. It constantly searches for the wake-up command to manage itself and simultaneously the other connected devices like smart lights, smart locks, smart doors, and so on. Thus, the devices which connect to Alexa to capture as well as respond to the commands can be used as a potential source of evidence.

Communications with Alexa along with the user's history data are stored in the SQLite database and web cache files [67–70]. The use and analysis of embedded files is also an effective method in the process of IoT forensics. Extracted data include device-related information, accounts, and networks. Detailed information about any IoT device, like the name of the device, its serial number, Wi-Fi and MAC address, and so on can be obtained by means of logical methods. The user's profile and settings provide information such as account details, Alexa associated devices, customer settings, user behaviour, user activity, and so on.

5.7.3 IoT IN A SMART HOME

In a smart home environment, smart devices like smart lights, smart switches, smart locks, smart cameras, smart televisions, and temperature sensors are used to improve quality of life as well as to reduce power consumption [71]. These home automation IoT devices employed in a smart home environment are also playing a big role in IoT forensics. These devices give the location of the suspect and their actions at the time of the incident.

On the other hand, these devices themselves are prone to cyber security attacks. For example, smart locks would be automatically unlocked if the mobile device of a person is within some range of distance. If this is known to an attacker, they can simply enter the home.

In case of a home or business environment, the attacker can disable the other smart home equipment, like cameras and alarms, or steal information to keep an eye on inhabitants of a smart home. Likewise, if a smart vehicle is attacked, it may lead to an accident.

5.8 SOLUTION METHODOLOGY PROPOSED

Various methods are used for the investigation of malicious activity by making use of IoT data. Some of them are discussed in this section.

5.8.1 MACHINE LEARNING ALGORITHMS

The malware detection in network traffic analysis is performed using machine learning algorithms [72]. An application and comparison of random Forest, Ada Boost, decision tree, neural network, SVM, and linear model data is performed. The complete forensics procedure is segregated in four modules. First is data collection and information production. The second module is about feature analysis and extraction. The third and fourth phases are designing of machine learning models and their analysis on various efficiency measures.

5.8.2 Public Digital Ledger

Hossain et al. make use of a distributed and decentralized public digital ledger framework for IoT forensics investigation purposes [66]. It stores information about each IoT device, its users, and data storage. A block represents a listing of transactions, which show the interaction with the IoT device. The list of such records is called blockchain. The ledger is available to all the stakeholders of the IoT device and they can retain a copy of the ledger. This helps to avoid the single centralized control. The interactions are signed and encrypted before storing in public blockchain. This ensures the authenticity of such data, which can be confidently used for forensic investigation.

5.9 OPPORTUNITIES AND FUTURE TECHNOLOGIES

This section presents the future opportunities that will enhance the field of IoT forensics.

5.9.1 Forensic Data Dependability

The potential evidence data collected should be dependable. There are sometimes privacy and security violations during the data collection phase. The creation of remote repositories that store all the data for future use will make the process less troublesome [73].

5.9.2 Models and Tools

Digital forensics methods are applied for IoT forensics that do not fit into the IoT framework. So, there is a requirement for new tools, methods, and technology to respond to the problems related to IoT forensics.

5.9.3 Smart Analysis and Presentation

There are many issues with the evidence data collected from many IoT sources. Presence of big data and the heterogeneity of data among multiple IoT devices make the analysis task cumbersome. So, new analysis methods and tools should be developed, which can use aggregated information and ensure the data inferences are uncomplicated and reliable.

5.9.4 Resolving Legal Challenges

Due to the low memory capability of IoT devices, in the majority of devices, data is stored in the cloud. The challenges associated with the storage and extraction of data in the cloud also have some legal issues. The data may be stored at some cross-border data store. So, extraction of that data has to go through the legal challenges. The use of cloud storage is necessary in IoT devices, so it should be handled carefully for successful and timely application of IoT forensics [74].

5.9.5 SMART FORENSICS FOR IoT

The Internet has enabled IoT devices to produce a huge number of records. The data collection process for forensics investigation faces many challenges due to heterogeneous data generated from different IoT devices and many others, as discussed in the previous section. The IoT forensics approach can be improved by making it smart. There is a solution for automated collection of data generated by IoT devices. Automation will help reduce overheads associated with data collection from a large number of potential evidence sources. It will also authenticate the forensic soundness of the data collection method. The data collection process may also start when there is some abnormality observed during the real-time working of the application. How to identify the malicious activity going on is still a question for research.

5.9.6 EMERGING TECHNOLOGIES FOR IoT

The role of big data is increasing in IoT forensics. Along with handling large quantities of heterogeneous data, it should also work without compromising on efficiency. Virtualization is used in distributed systems, mainly in the cloud environment. Virtual machines are popular for performing complex tasks that are executed at multiple points such as in the application, hardware, operating system, storage, and networks. The new technology that is replacing virtualization is containerization, which shares a single kernel among multiple applications, preferably on the same operating system [75]. In case of big data, containerization performs more quickly, is secure, and removes the latency problem. The data generated from IoT devices can be highly dimensional, which increases the complexity. Graph data structures can be used to analyse big data in different graph formats.

5.10 CONCLUSION

This book chapter presents a detailed study on the role and importance of large-scale data originated from IoT devices in the upcoming field of IoT forensics. Although it is always beneficial to have some insight from a large amount of data available, it is equally difficult to extract the relevant information from that large set. Moreover, the existing algorithms were designed for a comparatively smaller set of data. Hence, existing popular algorithms must be revised to enable the handling and processing of big data. Several challenges, including privacy and security of data, are discussed. IoT devices have been playing an important role for the acquisition of digital evidence. Based on the specific considerations and challenges associated with IoT devices, new digital forensics techniques must be designed. Future researchers need to address the challenges and suggest solutions for them. By proposing solutions for the existing concerns in IoT forensics, it may open a vast amount of opportunities in the field of crime investigation.

REFERENCES

[1] R. Hegarty, D. Lamb, and A. Attwood, *Digital evidence challenges in the internet of things*. In *Proceedings of the Tenth International Network Conference*, Plymouth, UK, pp. 163–172, 2014.

[2] E. Oriwoh and P. Sant, *The forensics edge management system: a concept and design*. In *UIC/ATC*, Vietri sul Mare, Italy. IEEE, 2013.

[3] L. Daniel and L. S. Daniel, *Digital Forensics for Legal Professionals: Understanding Digital Evidence from the Warrant to the Courtroom*. Syngress, Waltham, 2012.

[4] A. Gani et al., A survey on indexing techniques for big data: taxonomy and performance evaluation. *Knowledge and Information Systems* 46.2 (2016): 241–284.

[5] C. Perera, R. Ranjan, and L. Wang, Big data privacy in Internet of Things era. *Internet of Things Magazine*, pp. 32–39, 2015.

[6] K. S. Sarma and M. Raghupathi, Security issues of Big Data in IoT based applications. *International Journal of Pure and Applied Mathematics* 118.14 (2018): 221–227.

[7] Y.-S. Kang et al., MongoDB-based repository design for IoT generated RFID/sensor big data. *IEEE Sensors Journal* 16.2 (2016): 485–497.

[8] A. F. A. Rahman, M. Daud, and M. Z. Mohamad, *Securing sensor to cloud ecosystem using internet of things (IoT) security framework*. In *Proceedings of the International Conference on Internet of Things and Cloud Computing*. ACM, Cambridge, UK, March 2016.

[9] H. Ye et al., *A survey of security and privacy in big data*. In *2016 16th International Symposium on Communications and Information Technologies (ISCIT)*. IEEE, Qingdao, China, September 26–28, 2016.

[10] I. A. T. Hashem et al., The rise of "big data" on cloud computing: review and open research issues. *Information Systems* 47 (2015): 98–115.

[11] M. Al Fahdi, N. L. Clarke, and S. M. Furnell, *Challenges to digital forensics: a survey of researchers & practitioners attitudes and opinions*. In *Information Security for South Africa*, Johannesburg, pp. 1–8. IEEE, 2013, doi: https://doi.org/10.1109/ISSA.2013.6641058.

[12] M. Taylor, J. Haggerty, D. Gresty, and R. Hegarty, Digital evidence in cloud computing systems. *Computer Law & Security Review* 26.3 (2010): 304–308.

[13] L. Atzori, A. Iera, and G. Morabito, The Internet of Things: a survey, *Computer Networks* 54 (2010): 2787–2805.

[14] R. Minerva, A. Biru, and D. Rotondi, *Towards a definition of the Internet of Things (IoT)*, IEEE Internet Initiative, 2015.

[15] J. Gubbi, R. Buyya, S. Marusic, and M. Palaniswami, Internet of Things (IoT): a vision, architectural elements, and future directions. *Future Generation Computer Systems* 29.7 (2013): 1645–1660.

[16] R. Khan, S. U. Khan, R. Zaheer, and S. Khan, *Future internet: the internet of things architecture, possible applications and key challenges*. In *2012 10th International Conference on Frontiers of Information Technology (FIT)*, Islamabad, Pakistan, pp. 257–260. IEEE, December 2012.

[17] P. H. Rughani, IoT evidence acquisition – issues and challenges. *Advances in Computational Sciences and Technology* 10.5 (2017): 1285–1293.

[18] F. Servida and E. Casey, IoT forensic challenges and opportunities for digital traces. *Digital Investigation* 28 (2019): S22–S29.

[19] C. Altheide and H. Carvey, *Digital Forensics with Open Source Tools*. Syngress, Burlington, MA, 2011.

[20] S. Alabdulsalam et al., Internet of Things forensics – challenges and a case study. In *IFIP International Conference on Digital Forensics*. Springer, Cham, 2018.

[21] K. Ruan et al., Cloud forensics. In *IFIP International Conference on Digital Forensics*. Springer, Berlin/Heidelberg, 2011.

[22] M. E. Alex and R. Kishore, Forensics framework for cloud computing. *Computers & Electrical Engineering* 60 (2017): 193–205.

[23] J. Dykstra and A. T. Sherman, *Understanding issues in cloud forensics: two hypothetical case studies*. UMBC Computer Science and Electrical Engineering Department, 2011.

[24] R. C. Joshi and E. S. Pilli, *Fundamentals of Network Forensics*. Springer, London, 2016.

[25] E. Casey, Network traffic as a source of evidence: tool strengths, weaknesses, and future needs. *Digital Investigation* 1.1 (2004): 28–43.

[26] L. Morrison et al., Forensic evaluation of an Amazon fire TV stick. In *IFIP International Conference on Digital Forensics*. Springer, Cham, 2017.

[27] A. Pichan, M. Lazarescu, and S. T. Soh, Cloud forensics: technical challenges, solutions and comparative analysis. *Digital Investigation* 13 (2015): 38–57.

[28] H. Guo, B. Jin, and T. Shang, *Forensic investigations in cloud environments*. In *2012 International Conference on Computer Science and Information Processing (CSIP)*, Xi'an, Shaanxi. IEEE, 2012.

[29] M. Damshenas et al., *Forensics investigation challenges in cloud computing environments*. In *2012 International Conference on Cyber Security, Cyber Warfare and Digital Forensic (CyberSec)*, Kuala Lumpur, Malaysia. IEEE, 2012.

[30] J. Gregorio, B. Alarcos, and A. Gardel, Forensic analysis of nucleus RTOS on MTK smartwatches. *Digital Investigation* 29 (2019): 55–66.

[31] Europol, Internet organised crime threat assessment (IOCTA 2017\). European Union Agency for Law Enforcement Cooperation (Europol), 2017.

[32] A. MacDermott, T. Baker, and Q. Shi, *IoT forensics: challenges for the IoA era*. In *2018 9th IFIP International Conference on New Technologies, Mobility and Security (NTMS)*, Paris, France. IEEE, 2018.

[33] B. Alessio et al., *On the integration of cloud computing and internet of things*. In *2014 International Conference on Future Internet of Things and Cloud (FiCloud)*, Barcelona, Spain. IEEE, 2014.

[34] Z. A. Baig et al., Future challenges for smart cities: cyber-security and digital forensics. *Digital Investigation* 22 (2017): 3–13.

[35] D. Reinsel, J. Gantz, and J. Rydning, The digitization of the world from edge to core. IDC White Paper, 2018.

[36] H. F. Atlam et al., Blockchain with Internet of Things: benefits, challenges, and future directions. *International Journal of Intelligent Systems and Applications* 10.6 (2018): 40–48.

[37] Open Web Application Security Project, *IoT attack surface areas project*, 2018. https://www.owasp.org/index.php/OWASP_Internet_of_Things_Project#tab=IoT_Attack_Surface_Areas, February 7, 2020.

[38] J. I. James and Y. Jang, *Practical and legal challenges of cloud investigations*. arXiv preprint arXiv:1502.01133, 2015.

[39] J. Qiu et al., A survey of machine learning for big data processing. *EURASIP Journal on Advances in Signal Processing* 2016.1 (2016): 67.

[40] A. Jadhav and L. Deshpande, A survey on approaches to efficient classification of data streams using concept drift. *IJARCSMS* 4.5 (2016): 137–141.

[41] J. Sun et al., Dynamic financial distress prediction with concept drift based on time weighting combined with Adaboost support vector machine ensemble. *Knowledge-Based Systems* 120 (2017): 4–14.

[42] K. Alrawashdeh and C. Purdy, *Toward an online anomaly intrusion detection system based on deep learning*. In *2016 15th IEEE International Conference on Machine Learning and Applications (ICMLA)*, Anaheim, CA. IEEE, 2016.

[43] C. Yin et al., A deep learning approach for intrusion detection using recurrent neural networks. *IEEE Access* 5 (2017): 21954–21961.

[44] C. Yin, Y. Zhu, J. Fei, and X. He, A deep learning approach for intrusion detection using recurrent neural networks. *IEEE Access*, 5 (2017), 21954–21961.

[45] G. Zhao, C. Zhang, and L. Zheng, *Intrusion detection using deep belief network and probabilistic neural network.* In *2017 IEEE International Conference on Computational Science and Engineering (CSE) and IEEE International Conference on Embedded and Ubiquitous Computing (EUC)*, Vol. 1. IEEE, Guangzhou, China, July 21–24, 2017.

[46] N. Koroniotis, N. Moustafa, and E. Sitnikova, Forensics and deep learning mechanisms for botnets in Internet of Things: a survey of challenges and solutions. *IEEE Access* 7 (2019): 61764–61785.

[47] W. Zang, P. Zhang, C. Zhou, and L. Guo, Comparative study between incremental and ensemble learning on data streams: case study. *Journal of Big Data* 1 (2014): 1–16.

[48] V. Timčenko and S. Gajin, Machine learning based network anomaly detection for IoT environments. *ICIST Proceedings* 1 (2018): 196–201.

[49] H. Liu and M. Cocea, *Granular Computing Based Machine Learning: A Big Data Processing Approach*, Vol. 35. Springer, Cham, 2017.

[50] H. Liu and M. Cocea, Granular computing based approach for classification towards reduction of bias in ensemble learning. *Granular Computing* 2.3 (2017): 131–139.

[51] M. Antonelli, P. Ducange, B. Lazzerini, and F. Marcelloni, Multi-objective evolutionary design of granular rule-based classifiers. *Granular Computing* 1.1 (2016): 37–58.

[52] A. Skowron, A. Jankowski, and S. Dutta. Interactive granular computing. *Granular Computing* 1.2 (2016): 95–113.

[53] M. Pawlicki, M. Choraś, and R. Kozik, *Recent granular computing implementations and its feasibility in cybersecurity domain.* In *Proceedings of the 13th International Conference on Availability, Reliability and Security*, Hamburg, Germany, 2018.

[54] C. Meffert, D. Clark, I. Baggili, and F. Breitinger, *Forensic state acquisition from internet of things (FSAIoT): a general framework and practical approach for IoT forensics through IoT device state acquisition.* In *International Conference on Availability, Reliability and Security (ARES)*. ACM, Reggio Calabria Italy, August 2017.

[55] S. Perumal, N. M. Norwawi, and V. Raman, *Internet of Things (IoT) digital forensic investigation model: top-down forensic approach methodology.* In *Proceedings of the 5th International Conference on Digital Information Processing and Communications*, Sierre, Switzerland, pp. 19–23, 2015.

[56] V. R. Kebande and I. Ray, *A generic digital forensic investigation framework for Internet of Things (IoT).* In *Proceedings of the 4th International Conference on Future Internet of Things and Cloud*, Vienna, Austria, 2016, pp. 356–362.

[57] A. Valjarevic and H. S. Venter, A comprehensive and harmonized digital forensic investigation process model. *Journal of Forensic Sciences* 6.6 (2015): 1467–1483.

[58] R. McKemmish, When is digital evidence forensically sound? In I. Ray and S. Shenoi, eds., *Advances in Digital Forensics IV, The International Federation for Information Processing*, pp. 3–15: Boston: Springer, 2008.

[59] E. Casey, What does "forensically sound" really mean? *Digital Investigation* 4.2 (2007): 49–50.

[60] E. Oriwoh, D. Jazani, E. Epiphaniou, and P. Sant, *Internet of things forensics: challenges and approaches.* In *9th IEEE International Conference on Collaborative Computing: Networking, Applications, and Worksharing*, Austin, TX, pp. 608–615, 2015.

[61] I. Yaqoob, I. A. T. Hashem, A. Ahmed, S. M. A. Kazmi, and C. S. Hong. Internet of things forensics: recent advances, taxonomy, requirements, and open challenges. *Future Generation Computer Systems* 92 (2019): 265–275.

[62] N. E. Oweis, C. A. Aracenay, W. George, M. Oweis, H. Sorri, and V. Sansal, *Internet of Things: overview, sources, applications, and challenges.* In *Proceedings of the Second International Afro-European Conference for Industrial Advancement AECIA 2015*, Villejuif, France, pp. 57–67, 2015.

[63] S. Alabdulsalam, K. Schaefer, T. Kechadi, and N. A. Le-Khac. Internet of Things forensics – challenges and a case study. In *IFIP International Conference on Digital Forensics*, pp. 35–48. Springer, Cham, 2018.

[64] S. Sangeetha and G. SudhaSadasivam. Privacy of big data: a review. In *Handbook of Big Data and IoT Security*, pp. 5–23. Springer, Cham, 2019.

[65] A. Nieto, R. Roman, and J. Lopez, Digital witness: safeguarding digital evidence by using secure architectures in personal devices, *IEEE Network* 30.6 (2016): 34–41.

[66] M. Hossain, Y. Karim, and R. Hasan, *FIF-IoT: a forensic investigation framework for IoT using a public digital ledger.* In *2018 IEEE International Congress on Internet of Things (ICIOT)*, San Francisco, CA, pp. 33–40. IEEE, July 2018.

[67] N. Chavez, Arkansas judge drops murder charge in Amazon echo case, 2017. [Online]. Available: https://edition.cnn.com/2017/11/30/us/amazon-echo-arkansas-murder-case-dismissed/index.html.

[68] H. Chung, J. Park, and S. Lee, Digital forensic approaches for Amazon Alexa ecosystem. *Digital Investigation* 22 (2017): S15–S25. [Online]. Available: http://www.sciencedirect.com/science/article/pii/S1742287617301974

[69] S. Zawoad and R. Hasan, *FAIoT: towards building a forensics aware eco system for the Internet of Things.* In *Proceedings of the 2015 IEEE International Conference on Services Computing*, New York City, NY, USA, June 27–July 2, 2015, pp. 279–284.

[70] S. Li, K.-K. R. Choo, Q. Sun, W. J. Buchanan, and J. Cao. IoT forensics: Amazon Echo as a use case. *IEEE Internet of Things Journal* 6.4 (2019): 6487–6497.

[71] T. Zia, P. Liu, and W. Han, *Application specific digital forensics investigative model in internet of things (IoT).* In *International Conference on Availability, Reliability and Security (ARES)*. ACM, Reggio Calabria, Italy, 2017.

[72] G. S. Chhabra, V. P. Singh, and M. Singh. Cyber forensics framework for big data analytics in IoT environment using machine learning. *Multimedia Tools and Applications* 79 (2020): 15881–15900.

[73] M. Harbawi, The Internet of Things forensics: opportunities and trends, Unpublished.

[74] M. Bellare and P. Rogaway, *Provably secure session key distribution: the three party case.* In *Proceedings of the 27th Annual ACM Symposium on Theory of Computing*, Las Vegas, NV, pp. 57–66, 1995.

[75] M. Bellare, D. Pointcheval, and P. Rogaway, Authenticated key exchange secure against dictionary attacks. In B. Preneel, ed., *Advances in Cryptology, EUROCRYPT 2000*, Lecture Notes in Computer Science, pp. 139–155. Springer, Berlin/Heidelberg, 2000.

6 Integration of IoT and Big Data in the Field of Entertainment for Recommendation System

Yihang Liu, Wanying Dou, Zixuan Liu, Emanuel Szarek, and Sujatha Krishnamoorthy
Wenzhou Kean University, China

6.1 INTRODUCTION

What does the cellular life form have in common with the Internet of Things (IoT)? More than you may think. We all know that humans are composed of many different cells, working together to form our basic biological structure. With our technological advancements has come the ability to see ourselves at a much more intricate level. We have come closer to understanding that, as Barabasi and Oltvai [2] state, "alternatively to randomness the broad stage of internal order determines the cell molecular organization". An entire body of tiny organisms creates this massive web of structure and stability; they all work harmoniously for everything to continue to live. The IoT is no different. It is heading in a direction of connectedness that many would never even have imagined, with a goal that, as Alberti and Singh [1] state, objects in the environment will be sensed by network objects and will try to communicate with each other with the help of the Internet. The bountiful opportunities the Internet has given us can be seen by the massive amount of data that is being recorded, stored, and processed. Just like cells, the IoT is about gathering information from its surroundings to constantly learn, grow, and adapt. The basic and almost apparent way our lives are changing is providing marketers with the power to target individuals within a specific group. As we deal with large amounts of data, the information is properly organized with machine learning algorithms and analytics. Some platforms like social media and Twitter, Facebook, and networks permit organizations to instantly target one-on-one based on elaborated specifications that go beyond competition, age, gender, or cultural customs.

This invisible web spans across the globe, creating opportunities for human civilization to hopefully come closer together and feel as one. We use the IoT in different fields, such as medical education, science, biology, plant technology, and also in entertainment. This particular case study is to understand the application of the technology in the field of entertainment. As we know, Korean and Chinese dramas have lot of serials and movies online and there are various age groups of people who are fond of these movies and watch them online. Different researchers have found what enhances the recommendations for the user to watch next. The suggestions are based on the likes and interests of the user who watches the video and this recommendation system helps us to minimize the search time.

This case study is to help us better understand some of the technology that Netflix has used in order to recommend us shows and movies that we did not even know we wanted to watch. In the most basic form, Netflix started off with a recommendation system that comprised two different types, a recommendation based on a review system and a cooperative filtering system. Netflix would also add more algorithms, such as interleaving and context awareness, working in sync to create an ever-changing scheme.

6.2 BACKGROUND

Recommendation systems can be created in different ways, either text-based or review-based, as summarized by Li Chen [5]. There are many research directions in the recommendation systems. Figure 6.1 details a full picture of the analysis carried out with respect to reviews, but in general the recommendations are kernel-based, network-based, content-based, time-based, and so on. In this proposed work, we concentrate on Netflix and the algorithms used in the recommendation system.

Netflix wanted to make its algorithm even better and created a competition to see if anyone could create a new algorithm that would improve their current movie rating predictions. However, before Netflix was able to employ the winning team's algorithm, they were faced with the problem of speed. This did not hold back Netflix and they were quickly able to move past this issue. Netflix is a company that handles an immense amount of data and this ended up being very beneficial for everyone. We

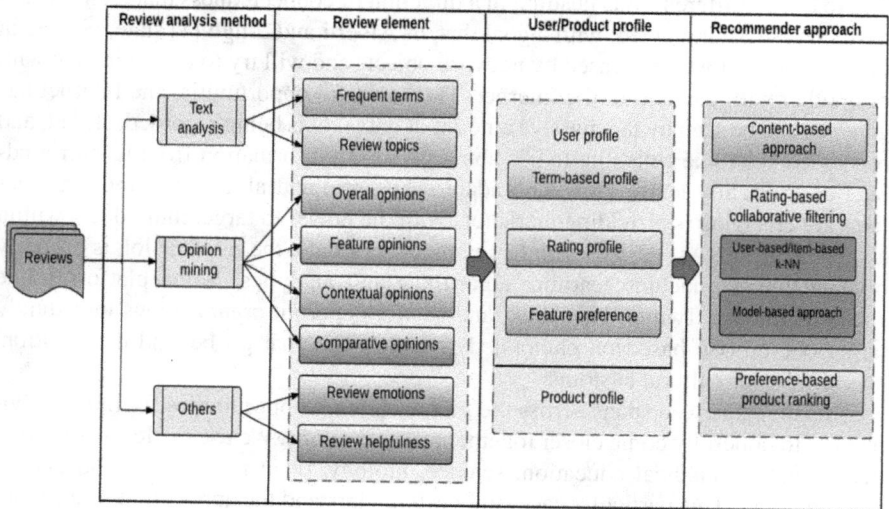

FIGURE 6.1 Investigation directions for review-based recommendation systems.

will look at why investing so much time and money into these features is and was so important.

Data science has become a sort of buzzword today. Many who work in the technology field understand the importance of data and why it is a currency that will end up rivalling oil. Khanduja et al. [12] stated that a survey by SAS confirms that more than 70% of companies trust that data science and big data analytics play a very important role. In today's market, most of the leading technology giants are most likely using some sort of data science algorithms to stay ahead of the competition. Netflix tapped into the vast power of data analytics while it was still in the early stages and since then has gone on to develop sophisticated algorithms for the company.

6.3 ANALYSIS AND ALGORITHMS

Netflix started as a digital optical disk rental company that slowly transformed over to online streaming in order to reduce postal losses [19, Para. 2]. During their switch to a streaming platform, Netflix would begin their journey towards becoming a streaming giant. Even during the early days of Netflix, they began investing time and money into newer and more efficient algorithms for their recommendation system. These recommendation systems would consider a whole host of data points. Almost anything the user did, Netflix was taking note of and using it to further develop their technology. An online article titled *You're not paranoid. Everyone is tracking you wherever you go, even Netflix* talks about how the author did not realize it, but Netflix was tracking his physical activity through his phone. It was an attempt to help make the streaming and buffering better for users who were on the go [9, Para. 2]. In general, it is assumed that the data analytics give better quantitative and qualitative information to enhance the business and services provided to the user. Nowadays, artificial intelligence (AI), business intelligence, and data science are becoming more and more common in most companies. One of the very main representatives of AI and film data analytics is at Netflix. Their entertainment service uses 33 million different versions of Netflix attributes assigned to their users, which are always called the Kissmetrics.

In 2017, to improve the efficiency of machine learning algorithms, Netflix no longer relied on the five-star rating, but used a combination of the "thumb system" and "percent match system". The percentage replaced the original star rating to represent Netflix's recommendation level for content. The reason was that the star rating method was too complex, and the active participation of users was not high. In addition, the five-star rating could not accurately reflect the preferences of users, while the behaviour of users when watching and their subsequent star rating could not be completely consistent. By contrast, the thumb system approach was considered more simple, as it only needs the user to express their view by selecting the thumb up or thumb down. It increased the number of users giving feedback and improved the efficiency of the algorithm.

Netflix's recommendation system uses a collection of recommendation formulae, the almost midpoint of which are individualized video rank (PVR) and Top-N video rank. The PVR algorithm is widely used in the Netflix recommendation system.

It provides personalized recommendation results for each user based on the movie type and arranges the whole type of reference according to individual liking. The objective of the Top-N algorithm is to provide insight into the most promising movies for users. Top-N is equivalent to an extension of the PVR algorithm. PVR ranks each recommendation, while Top-N finds the most relevant target from all sequences. Offline experience helps Netflix drive innovation. Based on the business model of paid membership, how to improve the retention rate of users becomes a good evaluation index of Netflix's recommendation system. The key to improving the retention rate of users lies in how to quickly meet the changing needs of the audience. As shown in Figure 6.2, the application of offline experience enables Netflix to continuously optimize relevant indicators through a large number of A/B tests. "Offline experiments allow us to repeat rapidly on algorithmic rule model, and to trim the user variants that we use in effective A/B venture" [4, 12]. Offline experience assists Netflix in driving innovation. Some algorithms are formed even with offline experience.

To begin, let us take a closer look at the basic filtering systems that Netflix used, content-based filtering and collaborative filtering. Content-based filtering uses what is already known about the product while also taking into consideration what the viewer is looking at [19]. For example, if you watch something that has sci-fi in the genre, it will recommend to you similar shows and movies that have sci-fi in the genre. The benefit of using this filter is that it can be more robust against quality bias and when the user is first starting out [14, Para. 10]. The negatives are that this filter only has the original item to work with and this can end up limiting the scope of recommendations, as well as suggesting items that have low ratings [14, Para. 10]. Collaborative filtering will give recommendations based on what similar users are watching. The benefit of using this type filtering is that it is "…'self-generating', users make the information for you naturally as they exchange information with items" [14, Para.7]. This also provides users with recommended shows and movies they may never have even thought about. Unfortunately, just like content-based filtering, collaborative filtering has issues when it comes to new users and there is nothing to work with. Instead of having one or the other, Netflix uses what is known as a hybrid recommendation system, which uses a mix of the two methods [19, Para. 7].

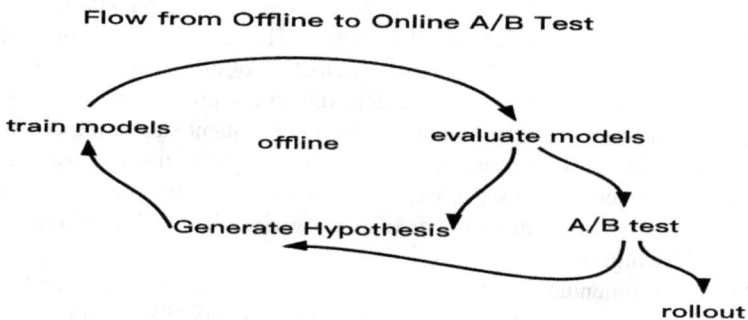

FIGURE 6.2   Offline experiment analysis.

The collaborative and the content-based filters are the two filters used in general in Netflix's recommendation system. These filters may use the text analysis or the content-based analysis in the filters. Figure 6.3 shows the idea of the filtering procedure happening in these methods.

Before going into case details, it is important to understand some of the basic concepts that will be used. Some of the key features were, K-nearest neighbour (KNN), singular value decomposition (SVD) and restricted Boltzmann. KNN is an algorithm that does what the name implies: it assumes similar things exist near each other [11]. Figure 6.4 shows the KNN visual with Boltzmann and KNN algorithm.

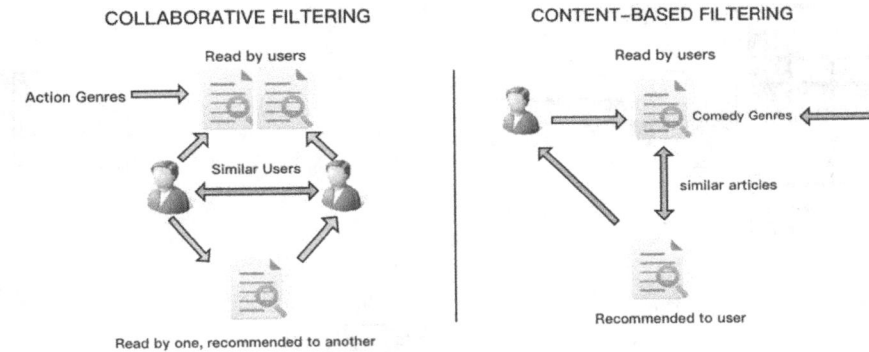

FIGURE 6.3 Collaborative filtering versus content-based filtering [19].

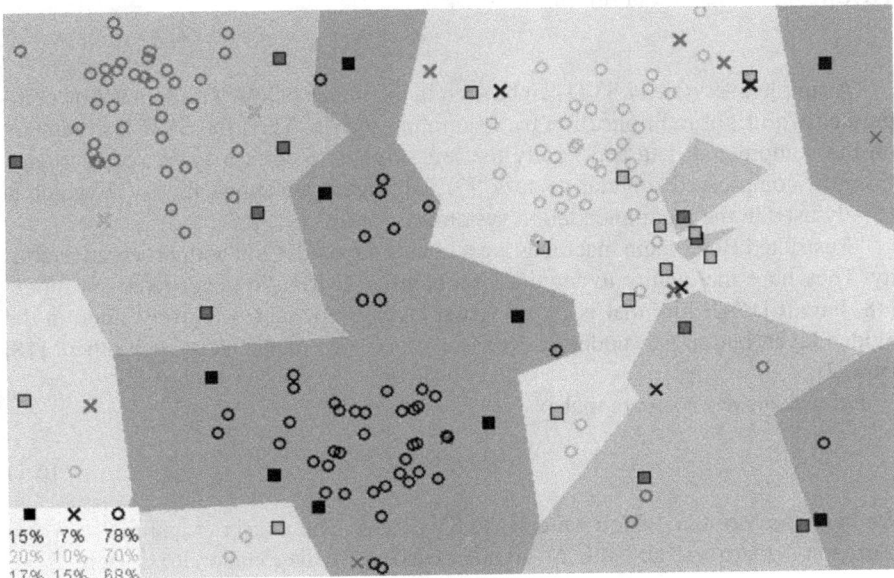

FIGURE 6.4 KNN – visual [11].

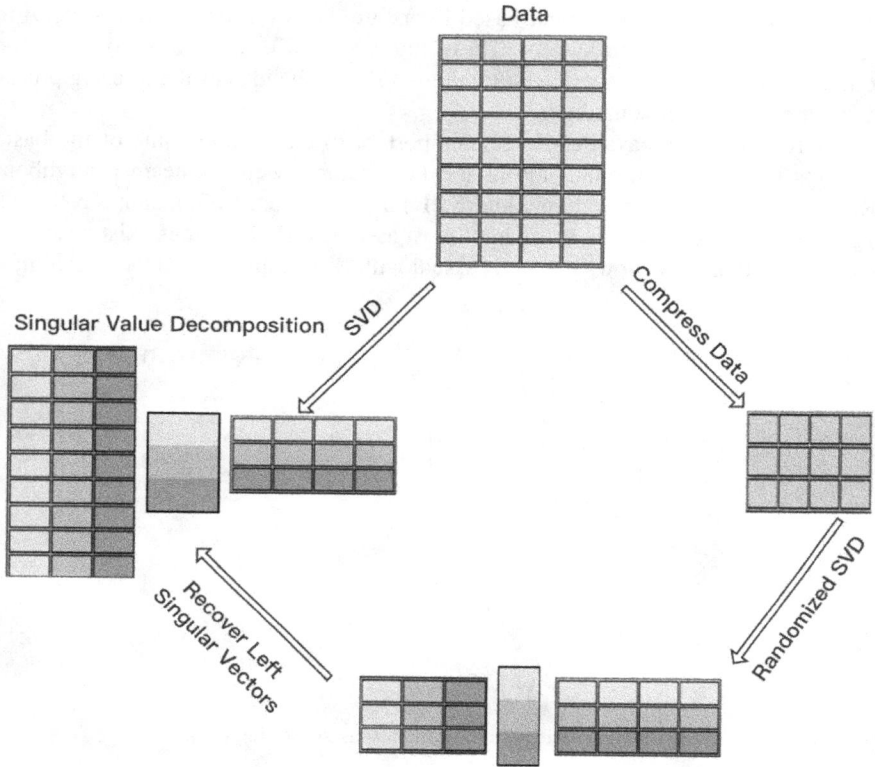

FIGURE 6.5 Overview of structure of SVD.

"A simple overview of SVD can be seen in the image below. The data is first compressed with right multiplication by a sampling matrix. Then the SVD is computed on the compressed data and finally the left singular vectors may be reconstructed from the compressed singular vectors" [7, 13]. Figure 6.5 shows the SVD, which is largely used in the recommendation system.

"Restricted Boltzmann machines are a double-layered ANN with procreative ability. They have the quality to acquire a probability distribution over its set of input" [18, Para. 6]. "All link that is in the visual layer is connected to every node in the hidden layer, but no two nodes in the same group are connected to each other" [18, Para. 7].

Let us consider a square matrix

$$A = U\Sigma V^T \tag{6.1}$$

We have the formula, which we call the SVD of A. The matrix U and matrix V are orthogonal matrices, and the Σ, sigma matrix is a diagonal matrix. Here is an example of the SVD algorithm illustrated with an example. Suppose we have a square matrix that is compressed by the original data matrix:

$$A^T A = V\Sigma^T U^T U\Sigma V^T \tag{6.2}$$

$$CV = U\Sigma \tag{6.3}$$

Matrix U is an orthogonal matrix, hence,

$$A^T A = V\Sigma^T U^T U\Sigma V^T = V\Sigma^T \Sigma V^T \tag{6.4}$$

Then, after getting the values of $A^T A$ we use A to get a new matrix which we gave as C using Equation (6.5)

$$C = \begin{bmatrix} c_{11} & c_{12} \\ c_{21} & c_{22} \end{bmatrix} \tag{6.5}$$

To calculate the eigenvalues of the matrix C, we give the eigenvalues as the symbol λ, then we use the formula:

$$\det(C - \lambda I) = 0 \tag{6.6}$$

In order to calculate the eigenvalue, we calculate the eigenvector V_n according to each eigenvalue, that is, the one of null spaces. It is important to state here that because the matrix V is an orthogonal matrix, we have to obtain the unit vector, which we give as V_n of each vector we just calculated by using the formula in Equation (6.7). (It is more mathematics to call the singular values instead of eigenvalues in this problem):

$$\hat{V}_n = \frac{1}{||V_n||} V_n \tag{6.7}$$

in which the $||V_n||$ means the length of the V_n.

Suppose the null spaces are obtained as shown in the below Equation (6.8)

$$\hat{V}_n = \begin{bmatrix} \hat{V}_{n11} \\ \hat{V}_{n21} \end{bmatrix} \tag{6.8}$$

These vectors are used to construct the matrix V (actually, V is the eigenvector matrix of $A^T A$):

$$V = \begin{bmatrix} \hat{V}_{1_{11}} & \hat{V}_{2_{11}} \\ \hat{V}_{1_{11}} & \hat{V}_{2_{11}} \end{bmatrix} \tag{6.9}$$

At the same time, construct the Σ by using the eigenvalues:

$$\Sigma = \begin{bmatrix} \sigma_1 & 0 \\ 0 & \sigma_2 \end{bmatrix} \tag{6.10}$$

After completing the above steps, further steps include two different methods for obtaining the matrix U. The first one is by using the formula given below:

$$CV = U\Sigma$$

The second method is to find the similarity using the same process again but obtaining the AA^T:

$$AA^T = U\Sigma^T V^T V\Sigma U^T = U\Sigma^T \Sigma U^T \tag{6.11}$$

As the matrix U is the eigenvector matrix of AA^T, we could repeat the same process introduced above to get the matrix U.

Finally, calculate the value of A using Equation (6.12):

$$A = U\Sigma V^T = \begin{bmatrix} \hat{U}_1 & \hat{U}_2 \end{bmatrix} \begin{bmatrix} \sigma_1 & 0 \\ 0 & \sigma_2 \end{bmatrix} \begin{bmatrix} \hat{V}_1^T \\ \hat{V}_2^T \end{bmatrix} \tag{6.12}$$

For the applied statistic, it is important to select appropriate samples to estimate the properties of the population. When applying the SVD algorithm, the first step is to compress a simple matrix from a large population data set. After we obtain the SVD, we need to reconstruct the results to the dimension that corresponds to the original matrix, in the case of a single prediction.

The main reason that the remote analysis is important when designing a sophisticated recommendation system is that each user has their own environment. Asking users to test the system in a particular lab will give the designers accurate data, while letting the users use the system in their own environment may present another magic phenomenon. For instance, if the user is now, after searching in the system, interested in long and romantic movies, the user may be very happy if the system next time could recommend some soap operas. The core thing there is the analysis of the series of actions that the user performed in the last period. After the algorithm filters and gives weighted values for each item, the system could generate some useful choices that the user may perform in the next stage.

It was as early as 2006 that Netflix realized there had to be something better than what their current algorithm was doing. The algorithm they were running was known as CineMatch and "Considering to Netflix, these suggestions were accurate inside half a star 75 percent of the time, and half of Netflix gave the five-star rating for their rental CineMatch-recommended movies gave them a five-star rating" [20]. The challenge was able to work quite well due to a couple of factors that were determined by Netflix. The first was giving users a large data set to work with. The data set consisted of more than 100 million movie recommendations, all anonymous [13, 1]. "In the set of data is about 48,000,189 users and 17,770 movies were recommended. Thus, the competitors had the help of a strong data set as well as luck." It was lucky Netflix decided to pick 10% because "small deviation in the figure may end up with competition to be too easy or any small deviation from this number, would have made the competition either too easy or impossibly difficult" [13, 9]. Figure 6.6 was while the challenge was still going on. However, based on the number of submissions Netflix was receiving, we can see why this was the case. Figure 6.6 indicates the "Additional projection with acknowledged methods (according to statement made on the Netflix Prize Forum) are also shown. These let in the per-movie mean and the per-user common rating" ([3], Figure 6.6).

There was also a great side effect from the competition. Creating a competition at this scale led the whole world's researchers to contribute to work on an individual problem. Everyone was united and collaborating to help each other try and figure this problem out. Netflix was only looking for an algorithm but, in the process, "to get connected with number of users with the movies they like and it is trying to do it with conceptual foundations of culture" [10, 118].

The use of new services brings a large amount of potentially valuable data for Netflix. How to fully mine this data has become the focus of their recommendation algorithm. Moreover, with the development of its business, Netflix has realized the transformation from a DVD leasing company to a large enterprise that can provide video streaming services to the world. This not only changes the way its users interact with the system but also changes their focus on recommendation algorithms. The

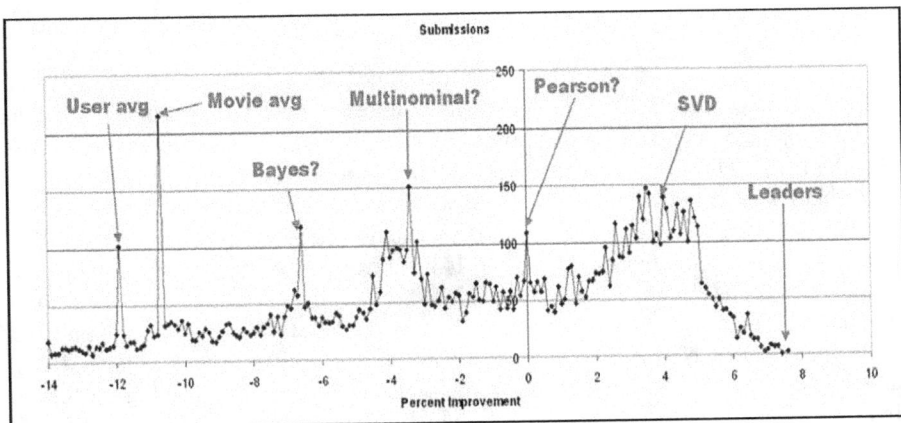

FIGURE 6.6   Techniques used for Netflix prize challenge.

most obvious change is the change of user feedback data type. Among the traditional DVD leasing services, the amount of information feedback provided by users is limited and inactive. It takes a long time for users to give feedback to the company after choosing a movie and watching it, and in this process, the company cannot receive any feedback from users, so it is difficult to know in time which movies are loved by users. With the publication of real-time streaming services, Netflix can at any time, through statistics, obtain data about the user watching the film, such as whether the user has seen the whole film, whether to recommend the film to others, and so on.

6.4   CASE STUDY

After three years, a team known as BellKor gave Netflix what they were looking for. The accuracy of CineMatch went up by 10.06% with an RMSE of 0.8712 [19, Para. 8]. The KNN algorithm was utilized for post-processing the data, then the SVD helped with best magnitude embedding to its users, and RBM increased the ability of the collaborative filtering model, thus creating Netflix's newest filter [19, Para. 8 and 9]. Netflix did struggle a little with implementation because, unlike the test data, Netflix's actual database contained many more data points. They also had reviews constantly being added to the system, unlike in the static model that was used during the competition. A research paper that was replicating the algorithms to see how they worked found that for speed computation "…gradient descent methods (stopping when RMSE on the probe set is minimized) worked effectively" [8, 226].

Content with their new algorithm, Netflix looked for more ways to improve its product. Netflix was running many different ranking algorithms that were all running simultaneously to try and provide suggestions that the viewer may like. However, having all these ranking algorithms run at the same time was hard and the speed was slow. In order to both improve the personalized recommendations and have high performance, Netflix innovated the algorithmic process of its current algorithms with what is known as interleaving.

One final key ingredient of Netflix's recommendation is its context awareness. This concept considers the following two categories: declared without conflict in position, spoken language, day and time, along with the device and inferred patterns [19, Para. 13]. Thus, the show that was watched at midnight during March madness and whether the user binged on it impacts what Netflix will recommend. This is done by taking "…the input as a sequence of user-actions and performs predictions that output the next set of actions" [19, Para. 14]. An example of this feature can be seen with Gru4Rec, as shown in the Figure 6.7. When Netflix started off, it had a recommendation system that catered to their audiences but Netflix knew it was not enough. After many years, Netflix's recommendation system is now at the next level, being able to harness CineMatch's recommendations, and using interleaving to further improve personalization of these recommendations and context awareness to finish it all by adding importance to the recommendations.

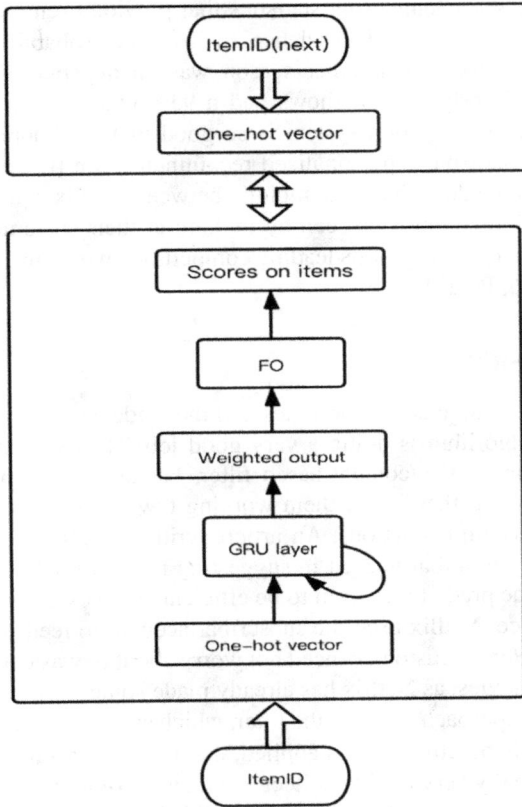

FIGURE 6.7 Example of context aware method using the sequence prediction using Gru4Rec.

6.5   DISCUSSION

The amount of time and effort that Netflix spent on its algorithms can be seen by how widely successful it has become. Their ability to use data has become so proficient that they made one of their biggest company decisions from it. When *House of Cards* was up for grabs, Netflix made sure to do whatever it took to grab the rights. They outbid competitors like HBO and AMC for a two-season contract worth over USD 100 million [17, Para. 30]. There were three key statistics that made Netflix so sure about this deal. One, there were a lot of people who ended up watching *The Social Network* from start to end. Two, there was a British version that was like *House of Cards* and that show had a very strong viewer base. Three, there was a correlation between people who watched the British version and those who also watched Kevin Spacey films and/or films directed by David Fincher. Now Netflix is reaping the benefits from its calculated investment.

Netflix also has some staggering statistics that provide even more proof of how well they have been able to data science. In 2017, the probability that a cable TV show would be renewed for another season was an abysmal 35% [6, Para. 3]. Meanwhile Netflix's original TV shows had a 93% chance of being renewed [6, Para. 3]. Their recommendation system is so good that a whopping 75% of what viewers watched was from a personalized recommendation [6, Para. 5]. This relationship can almost be described as symbiotic between Netflix and the viewer. Their customer satisfaction can also be seen by looking at their customer retention rate, 93%, which is higher than any of its leading competitors, with Amazon Prime at 75% and Hulu at 64% [6, Para. 1].

6.6 CONCLUSION

Netflix has been able to adapt to the times and the evidence is quite clear that their recommendation algorithm is doing a very good job. They started out with just a simple hybrid collaborative/content-based filter, but now have multiple high-end algorithms running together, all of them working towards giving Netflix's viewers the best possible recommendations. An article written in 2017 concluded that the recommendation system that is used in suggesting movies for Netflix users is very accurate and that the predictions seem to be efficient. Moreover, in order to improve the user's experience, Netflix used the subscriber account to feed suggestions to the queue of the user. Hence, customers need not worry about or waste their time in hunting for trailers or movies, as Netflix has already made some good recommendations. The most common approach is to use the filter, which sometimes omits the metadata. In cases when a collaborative filter is applied, the human-generated textual data can be used, which is really very hard to collect and may also end in an error. Sometimes the tags created by users and the comments and the thumbs-up can be used to generate suggestions in some of the very famous algorithms. Hence, a future direction for the researchers is to use the textual metadata. Some dimensions of the keywords can help to bring text data together and the computational model can be arrived at without any further errors in the recommended system.

REFERENCES

[1] Alberti, A. M. and Singh, D. "Internet of Things: Perspectives, Challenges and Opportunities." 2013. Retrieved from https://www.researchgate.net/publication/236656851_Internet_of_Things_Perspectives_Challenges_and_Opportunities.

[2] Barabasi, A.-L. and Oltvai, Z. "Network Biology: Understanding the Cell's Functional Organization." *Nature Reviews Genetics*, vol. 5, 2004, pp. 101–113. doi:10.1038/nrg1272.

[3] Bennett, J., Lanning, S., and Netflix, Netflix. *The Netflix Prize*, 2009.

[4] Gomez-Uribe, C. A. and Hunt, N. "The Netflix Recommender System: Algorithms, Business Value, and Innovation." *ACM Transactions on Management Information Systems*, vol. 6, no. 4, December 2015, pp. 13.1–13.9. doi: http://dx.doi.org/10.1145/2843948.

[5] Chen, L., Chen, G., and Wang, F. "Recommender Systems Based on User Reviews: The State of the Art." *User Modeling and User-Adapted Interaction*, vol. 25, no. 2, 2015, pp. 99–154.

[6] Dixon, M. "How Netflix Used Big Data and Analytics to Generate Billions." *Selerity, Scorpio Software Services Pty Ltd T/A Selerity*, October 18, 2019, seleritysas.com/blog/2019/04/05/how-netflix-used-big-data-and-analytics-to-generate-billions/.

[7] Erichson, N. B., et al. "Randomized Matrix Decompositions Using R." *Journal of Statistical Software*, vol. 89, no. 11, April 2018. doi:10.18637/jss.v089.i11.

[8] Feuerverger, A. et al. "Statistical Significance of the Netflix Challenge." *Statistical Science*, vol. 27, no. 2, 2012, pp. 202–231. JSTOR, www.jstor.org/stable/41714795. Accessed April 2, 2020.

[9] Gates, C. "Everyone Is Tracking You Wherever You Go, Even Netflix." *Digital Trends*, August 1, 2019, www.digitaltrends.com/home-theater/netflix-tracking-android-activity-data/.

[10] Hallinan, B. and Striphas, T. "Recommended for You: The Netflix Prize and the Production of Algorithmic Culture." *New Media & Society*, vol. 18, no. 1, 2014, pp. 117–137. doi: 10.1177/1461444814538646.

[11] Harrison, O. "Machine Learning Basics with the K-Nearest Neighbors Algorithm." *Medium, Towards Data Science*, July 14, 2019, towardsdatascience.com/machine-learning-basics-with-the-k-nearest-neighbors-algorithm-6a6e71d01761.

[12] Khanduja, A. "How Analytics Can Be a Game Changer: A Netflix Case Study." *Markivis, Markivis Pvt. Ltd*, February 25, 2020, www.markivis.com/insights/marketing-analytics/how-analytics-can-be-a-game-changer-a-netflix-case-study/.

[13] Koren, Y. "The BellKor Solution to the Netflix Grand Prize." *Netflixprize*, August 2009, www.netflixprize.com/assets/GrandPrize2009_BPC_BellKor.pdf.

[14] Lineberry, A. and Longo, C. "Creating a Hybrid Content-Collaborative Movie Recommender Using Deep Learning." *Medium, Towards Data Science*, September 11, 2018, towardsdatascience.com/creating-a-hybrid-content-collaborative-movie-recommender-using-deep-learning-cc8b431618af.

[15] Mandal, G., Diroma, F., and Jain, R. "Netflix: An In-Depth Study of Their Proactive & Adaptive Strategies to Drive Growth and Deal with Issues of Net-Neutrality & Digital Equity." *IRA-International Journal of Management & Social Sciences*, vol. 8, no. 2, 2017, pp. 152–161. doi: http://dx.doi.org/10.21013/jmss.v8.n2.p3, ISSN: 2455-226.

[16] Opportunities. Retrieved from https://www.researchgate.net/publication/236656851_Internet_of_Things_Perspectives_Challenges_and_Opportunities.

[17] Patel, N. "How Netflix Uses Analytics to Select Movies, Create Content, & Make Multimillion Dollar Decisions." *Neil Patel, I'm Kind of a Big Deal, LLC*, January 24, 2020, neilpatel.com/blog/how-netflix-uses-analytics/.

[18] Sharma, A. "Restricted Boltzmann Machines-Simplified." *Medium, Towards Data Science*, December 6, 2018, towardsdatascience.com/restricted-boltzmann-machines-simplified-eab1e5878976.

[19] DataFlair Team. "Data Science at Netflix – A Must Read Case Study for Aspiring Data Scientists." *DataFlair*, May 4, 2019, data-flair.training/blogs/data-science-at-netflix/.

[20] Wilson, T. V. and Crawford, S. "How Netflix Works." *HowStuffWorks*, May 14, 2007, electronics.howstuffworks.com/netflix2.htm.

7 Secure and Privacy Preserving Data Mining and Aggregation in IoT Applications

Vashi Dhankar and Anu Rathee
Maharaja Agrasen Institute of Technology, Delhi

7.1 INTRODUCTION

Every year, the area of IoT applications keeps on expanding. It has become an essential part of collecting and analysing data for robust healthcare systems, smart homes, and smart cities. This is being done at a rate never seen before. The unprecedented growth has also increased our capability of handling large volumes of high velocity, variety, and value data through big data ecosystems such as Hadoop. However, the essential need is to perform data mining, aggregation, and analytics on this data without having any privacy and security concerns. Currently, there are many privacy issues that come with mining and aggregation of data.

Privacy protection is intended to shield individuals' information and delicate data from public exposure while data and information mining take place. All the stakeholders are worried about data sharing and hence guard data from one another. For example, business collaborators working together on a platform or service would not want their sensitive information to be shared or leaked with each other while working together on a task. The IoT applications are profoundly flexible and different, which makes their needs very different. The commonly faced protection and privacy issues in the IoT environment are discussed in Section 7.2.

Data mining is done one step at a time and each part of the process requires its own method of security and privacy preservation. A structure for privacy preserving data mining (PPDM) is discussed in the third section. The framework divides the data mining process into three layers: the data collection layer (DCL), data pre-processing layer (DPL), and data output layer (DML). All the layers have their mechanism for privacy preservation. For example, randomization techniques are used in the DPL. Different approaches, such as personalized privacy and differential privacy schemes, are also discussed. All the methods with their core concepts, advantages, and disadvantages are compared with one another and their fields of application are discussed.

In IoT structures, an aggregator handles all activities for a collection of IoT devices. The aggregated data is then used as the input on which the analytics is done. Privacy has to be maintained with respect to content as well as context. In real-life scenarios, maintaining anonymity is also essential. There are many ways in which the data aggregation process can be made secure. In Section 7.4 of this chapter, we look at cryptographic techniques, data slicing methods, and evolutionary methods to obtain privacy preserving data aggregation. Homomorphic encryption and advanced encryption standard (AES) algorithms' working and application are discussed. Security analysis of the algorithm is done for eavesdropping, replay, manipulation, internal, collusion, and impersonation attacks. The performance is then evaluated considering the cost of encryption and decryption for all the phases such as data division, authentication, and aggregation. An evolutionary game-based model is also discussed. In this, the nodes are part of a community structure where behaviour of one user influences others in terms of cost, services, and utility. Data aggregation using data slicing methods is also mentioned in depth. Two methods, SMART and iPDA, are discussed in length with their advantages and disadvantages.

7.2 PRIVACY AND SECURITY CHALLENGES IN IoT APPLICATIONS

Before discussing the ways to ensure security and privacy for IoT applications, some of the challenges to security and privacy are discussed.

7.2.1 IDENTIFICATION

Profiling and authentication for IoT devices mean that we are partnering an identity, for example, a location with a person. In this context, there can be cases where a user's privacy is breached as some sensitive information may be given outside the user's personal sphere.

7.2.2 LOCALIZING AND TRACKING

Nowadays, there are many ways, such as GPS and IP addresses, which can assist with following a person's presence in both time and space. While it only gives a better user experience and enables more features, some users see it as an invasion of privacy, especially if the data is used inappropriately without their consent. Currently, the IoT faces challenges with how to deal with third-party data. There lies a huge challenge and moral dilemma balancing between the business interests of companies and privacy of the users.

7.2.3 LIFE CYCLE TRANSITIONS

It is necessary to keep track of all the updates available in an application's life cycle and how to apply them uniformly across distributed environments, whether the device is old or new. The update can also include a security patch which, if not applied to the device, can cause security issues in the whole network. This can prove to be a challenge in an industry based on applications, as discussed by Darmstadt [1].

7.2.4 SECURE DATA TRANSMISSION

For data transmission through open mediums, such as public mediums, it is necessary to follow safety measures. The information must be concealed to prevent unauthorized access and collection of information. The data must be protected from any internal as well as external adversary.

From surveys [2] and reviews [3] of privacy and security issues in IoT, it has been found that conventional safety efforts cannot be applied to the IoT. In the next section, an architecture is discussed that can manage the dynamic security needs for an integrated IoT environment.

7.3 SECURE AND PRIVACY PRESERVING DATA MINING TECHNIQUES

Since the process of data mining has many steps, there are different privacy preservation schemes and paradigms on each level of data mining. The main guiding idea for the PPDM is inspired by Zhang and Zhao [4]. The main layers that are considered in this chapter include the DCL and DPL. The DPL can also be considered as a sublayer. The final layer is the data mining output layer. All these layers are visible in Figure 7.1.

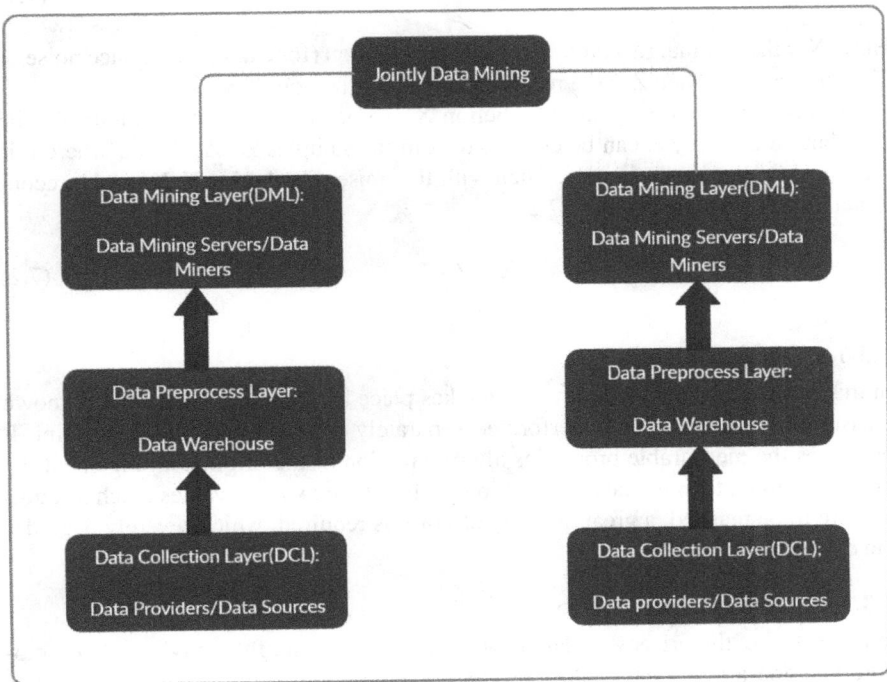

FIGURE 7.1 A PPDM framework.

At the first layer, data is collected from a huge number of resources. This is the raw, original data and contains users' private information. At the time of data collection, the privacy preserving techniques are utilized. This data is stored in data warehouses in the next layer before processing.

In the second layer (DPL), there are a number of data warehouse servers. Here, the raw data is stored and aggregated. This data may be aggregated as sum, average, or pre-computed using a method to ensure confidentiality. This makes the data aggregation and fusion process faster. The third layer is the data mining output layer (DML). Here, the results are obtained using some data mining algorithm. The job is to keep the data mining methods efficient while also incorporating security and privacy preservation. There can be other forms of data mining as well, such as collaborative data mining. Here, the database is shared, hence it is required to make sure that data sets owned by multiple parties do not reveal each other's information.

7.3.1 PRIVACY PRESERVING TECHNIQUES AT DATA COLLECTION LAYER

At this layer, there is a security and privacy concern if the sensitive data is collected by an untrustworthy collector. To prevent this, randomization is employed, which transforms the original data. The original data in its raw form is not stored as it will not be used further. Randomization can be described using the following equation.

$$Z = X + Y \tag{7.1}$$

Here, X = the original raw data. Y = noise distribution (for example, Laplace noise or Gaussian noise). Then Z = the result of the randomization of X and Y.

To reconstruct the original distribution X, we need an estimate of entity Z. The distribution of entity Z can be estimated from the samples $Z1, Z2, \ldots ZN$, where n is the total number in the sample. Then with the noise distribution Y, X can be reconstructed, as shown in Equation 7.2.

$$X = Z - Y \tag{7.2}$$

7.3.1.1 Additive Noise

In this method, the data randomization takes place by including noise with a known measurable dispersion. It is performed separately for each captured data point. It preserves the measurable properties after recreation of the original arrangement. Its usage is limited to aggregate distributions only. When extreme values (such as outliers) are to be masked, a great quantity of noise is required, which severely degrades the data utility.

7.3.1.2 Multiplicative Noise

In this scheme, the privacy of data is obtained in the form of the product of commotion data and a given data value. In this, the recreation of the first individual data value is tougher than in additive noise, thus making it a more secure privacy mechanism. It preserves the measurable properties after recreation of the original arrangement.

7.3.2 PRIVACY PRESERVING AT DATA PUBLISHING LAYER

Entities may seek to publish their data publicly for further research or analysis. However, malicious adversaries may attempt to de-anonymize or target record owners for malevolent purposes. There are several methods and privacy models for data anonymization. A sanitizing operation is applied on the data with the aim of preserving the record owner's identity. Some of these operations are:

7.3.2.1 Generalization

This method works for both categorical and numerical data [5–9]. Here, the actual value is substituted by a parent value, which is a general value. For example, if the total number of students in a class is 60, the value can be specified to be falling in an interval. For categorical data, a hierarchy is defined.

7.3.2.2 Suppression

This method is applied to a data set where either the column or row values are suppressed. The suppression is done by removal of some attributes. This helps in preventing the disclosure of any important information.

7.3.2.3 Anatomization

In many databases there are quasi-identifiers (QIDs). These are pseudo identifiers for a user. QIDs are not sensitive like usual IDs. They can be formed by one attribute or a set of attributes. In anatomization, the QIDs and sensitive attributes are divides into two separate relations. This helps in de-association of data from the QID. This does not change the value of the data.

These are the most commonly used schemes in the DPL. Based on these, the following data privacy models are defined:

7.3.2.4 K-Anonymity

K-anonymity uses generalization and suppression as its sanitization methods. It was proposed by Samarati and Sweeny [10, 11]. Anonymity is ensured by the presence of k–1 undistinguishable records for each record in a database. This set of k records is known as the identity grade. The attacker cannot identify a single record k and attack as there are similar k–1 records. It is a simple algorithm with a large amount of work done on the existing algorithms. The algorithm works best when the value of k is high. It assumes that each class element represents a distinctive value. The code characteristics play no role in anonymization, which can disclose information, especially if all records in a class have the same value for the sensitive attribute. There are other consequences of not taking sensitive attributes into consideration, such as de-anonymization of an entry, when a QID is associated with knowledge of the attribute and the database. Its application domain is wireless sensor networks [12], location-based services [13, 14], the cloud [15], and e-health [16].

7.3.2.5   L-Diversity

L-diversity builds upon the k-anonymity model by keeping a pre-requisite that each equal or comparable group must have L "well-represented" value for the sensitive attributes, hence it has similar generalization and suppression as its sanitization methods. For example:

$$-\sum_{t\in T} R\left(Id,t\right)\log\left(R\left(Id,t\right)\right) \geq \log\left(m\right) \tag{7.3}$$

Here, t = sensitive attribute (T) possible values and R(Id) = fraction of records forming one Id equivalence group, which have the value t for T attribute.

Entropy l-diversity can be stretched out to different characteristics. These sensitive attributes anatomize the data [16]. The variation in confidential data is also deliberated upon while anonymizing. This can be understood by an example similar to that proposed by Fung et al. [17]. Consider we have a data set where 93% of the entries have health insurance and 7% do not. An attacker seeks to find the group which does not have insurance, and has the original sensitive attribute knowledge. We will form its l-diverse group. The maximum entropy within the group will be achieved when 50% of the group have health insurance entries and the remaining 50% do not. Its application domain is in e-health [16] and location-based services [14, 18, 19].

7.3.2.6   Personalized Privacy

Personalized privacy is accomplished by making a scientific categorization tree utilizing generalization and also permitting the record proprietors to characterize a guarding node. Proprietors' security is penetrated if a malicious user is permitted to gather any secret incentive from the subtree of the guarding node with a likelihood (break likelihood) more prominent than a specific limit. In this, the proprietor can characterize their protection level. It maintains most utility while considering individual protection. However, it is hard to implement in practice. This is because it might be tough to approach and get records from owners. Also, there can be a tendency to overprotect data through general guarding nodes only. Its application domain is social networks and location-based services.

7.3.2.7   Differential Privacy

All algorithms in differential privacy rely on background knowledge. There are other algorithms which do not follow this paradigm, for example, the differential privacy paradigm which achieves data privacy through data perturbation. In these methods, a minuscule amount of noise is summed with the true data. Actual data is thus masked from the adversaries. The core idea of differential privacy is that addition or removal of one record from the database does not reveal any information to an adversary. The step-by-step working of the core idea of differential privacy is illustrated in Figure.7.2.

This means that your presence or absence in the database does not reveal or leak any information from the database. This achieves a strong sense of privacy.

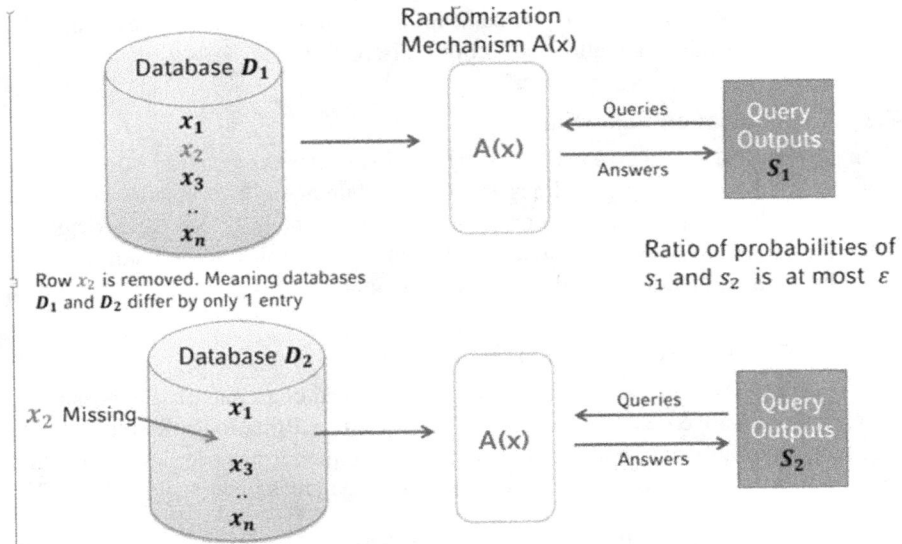

FIGURE 7.2 Differential privacy mechanisms.

7.3.2.8 ∈-Differential Privacy

∈-differential privacy ensures that a solitary record does not significantly influence (adjustable through value e) the result of the investigation of the data set. In this sense, an individual's protection will not be influenced by partaking in the information assortment, since it will not have a huge effect on the ultimate result. It ensures proper security and a strong protection loss metric. It also ensures that the participation of a single individual does not lead to a privacy breach greater than that obtained from the non-participation of the same individual. There is no experimental guide on setting it, as it strongly depends on the data set. Privacy guarantees can require heavy data perturbation for numerical data, leading to non-useful output. Its application domain includes e-health, smart meters [20], and location-based services. ∈-differential privacy uses a randomized mechanism A(x) for two databases, D_1 and D_2, that differ on at most one element, all output S range (A),

$$\frac{\Pr\left[A\left(D1\right) \in S\right]}{\Pr[A\left(D2\right) \in S}] \leq \exp(\in) \tag{7.4}$$

ε is privacy parameter called privacy budget or privacy level.

7.3.3 PRIVACY PRESERVING AT DATA MINING OUTPUT LAYER

The output of data mining techniques (for example, classifiers) is often made accessible through applications or interfaces. Malicious users can query these systems to infer sensitive information about the basic information. In these scenarios, either the data or the applications are altered to prevent disclosure.

7.3.3.1 Association Rule Hiding

Data is perturbed in order to avoid mining sensitive rules (association rule mining). Optimally, all non-touchy guidelines are mined, while no delicate standard is found. Numerous algorithms have been proposed but the problem is NP-hard, requiring heuristic arrangements that tend to also hide non-delicate rules (i.e., loss of information). Its application domain is the cloud and social networks [21].

7.3.3.2 Classifier Effectiveness Downgrading

Data is sanitized to reduce the classifiers' accuracy, and consequently the possibility of inferring sensitive data. Few standard classifiers use affiliation rule mining techniques as subroutines. Affiliation rule concealing strategies are additionally applied to minimize the viability of a classifier. Its application areas include cloud [22, 23].

7.3.3.3 Query Auditing and Inference Control

Query auditing and inference control consists of techniques to prevent information disclosure from sequences of aggregate data queries. In question derivation control, either the first information or the yield of the inquiry is bothered. In online question reviewing, at least one from a sequence is denied, whereas in offline query auditing, past query sequences are analysed to evaluate if the output breached privacy. This has the advantage that there is extensive research in the context of statistical database security, but it denies and blocks certain queries which can also reveal information. Its primary application area is e-health [24].

7.3.4 DISTRIBUTED PRIVACY

Numerous entities may try to mine worldwide bits of data as total insights, out of the combination of all parcelled information, without uncovering neighbourhood data from different elements. Semi-legitimate adversaries participating in distributed computation can try to gain extra knowledge from the shared data and, in such cases, deviate from the protocols. In this scenario, secure multiparty protocols are implemented to prevent disclosure of local data sets. Some of the protocols in it are:

7.3.4.1 One out of Two Oblivious Transfer

This is a secure protocol between two parties, where one message out of two messages is received and decrypted by the receiver, and the sender who has input the pair of messages is oblivious to which message was decrypted [25, 26]. The input messages are encrypted with the public keys sent by the receiver, and only one of the private keys for the decryption is held by this party. This convention has been summed up to the instance of k out of N members. It is used for secure and private exchange of information. Its applications are in the areas of e-health [27, 28] and the cloud [29].

TABLE 7.1
Protocols and their descriptions

Protocol	Description
Secure Sum	Obtains the aggregation from individual sites, without revealing such contributions to the next member entities [35,36].
Secure Set Union	A method to share different structures between member elements so as to make associations of sets, without revealing the proprietors of each set [36,37].
Scalar Product	Computes the scalar product between two parties without revealing the input vector to the other entity. A secure scalar product is of crucial importance, since many data mining problems can be reduced to the computation of the scalar product [38, 39].
Set Intersection	Computes the intersection of two sets, one from each participating party, without revealing any other information [40].

7.3.4.2 Homomorphic Encryption

Homomorphic encryption allows for algebraic computations on cipher text in a way that the deciphered result matches the result of the algebraic operation with the plaintext that originated the cipher text. It is used to secure data while stored, or in transactions while allowing for computation over encrypted data. Its applications are in the areas of e-health [30, 31], the cloud [32], and wireless sensor networks [33].

The next sets of protocols is used for a similar purpose, namely performing operations that are used as building blocks for distributed data mining techniques. These primitives limit the measure of data that is discharged to other participating entities. Most of their application is limited to the cloud [34]. They can be compared as shown in Table 7.1.

7.4 SECURITY ENSURING TECHNIQUES FOR PRIVACY PRESERVING DATA AGGREGATION

IoT devices are constrained with short battery power and also by the limited amount of available computational resources. This is where data aggregation comes into the picture. Using data aggregation, the network can be made energy efficient [41]. Data aggregation is the procedure where some nodes (or only one node) gather the results collected by each individual node. The node that collects the aggregated data is called the base station. This can also be a foreign node (handled by a user of the network) which is permitted to access the network. It is also referred to as the sink. The data aggregation process in the IoT is shown in Figure 7.3.

Generally, nodes in the IoT grid have sufficient memory to store the data generated by it or received from other nodes for only a short period of time. This data is then aggregated from different nodes and sent out as one whole aggregated result. In this scenario, the non-redundant data is not sent and the cost associated with it is nullified. This has a further benefit of increasing the network lifetime and also making it cost efficient.

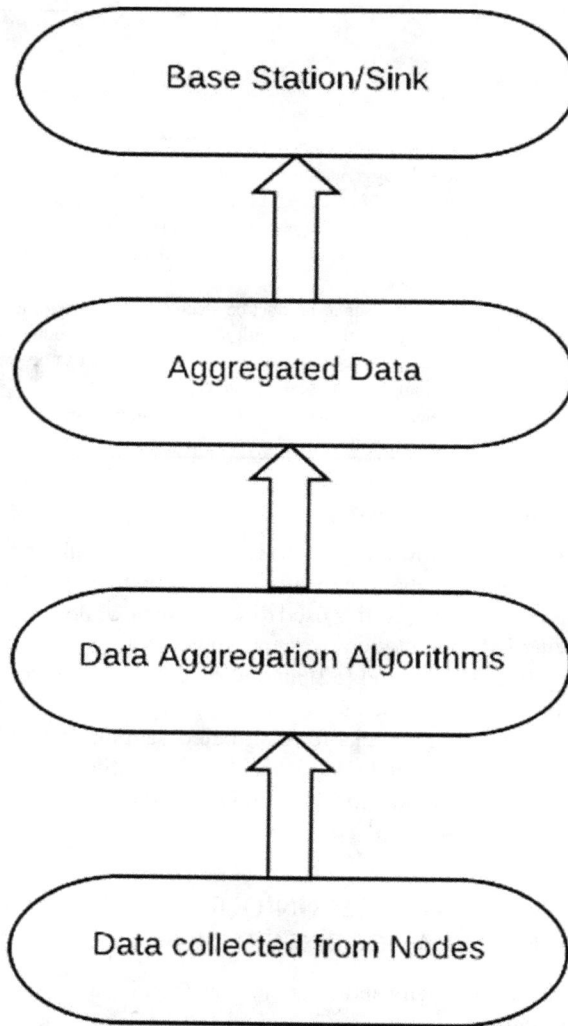

FIGURE 7.3 Architecture of data aggregation.

There are several algorithms which can be used to choose the most effective route to move data to the sink. The advantages of these methods are as follows:

- Data aggregation helps to achieve accuracy of information for the whole network. It is possible as it helps to improve the efficiency in the network.
- The unnecessary and redundant information coming from the network's nodes is decreased.
- As redundancy decreases, the traffic load decreases, which saves energy.

While there are many advantages of IoT, the data collected by the nodes should stay secure and there should not be a breach of privacy. Many of the IoT devices used in

smart homes, cities, and hospitals or having users' personal data can become pervasive. For example, smart monitors in hospitals can be used by attackers to eavesdrop
on real-time medical information about a patient, or a passive attack can also be
performed using only smart cameras and no other advanced tool, as the attacker can
carry out surveillance on the camera's user. Hence, to overcome these challenges,
many secure schemes have been proposed as follows:

7.4.1 Privacy Preservation Using Homomorphic Encryption and Advanced Encryption Standard (AES)

In this scheme it is considered that all the neighbouring IoT devices form a group and
they have both computation capacity and storage [42]. They have IDs, a common
secret key, and a private parameter τ. The aggregator is responsible for managing one
group. Only the aggregator and the device itself know the unique ID. The AES works
on the aggregator [43]. Using the proposed method, the aggregator can receive the
collected data from devices. The scheme has a total of five stages as follows:

7.4.1.1 Implementation of Homomorphic Encryption and AES Algorithm

Division of Data

- Assume that an aggregator covers an area with nodes $N = \{N1, N2, N3..., Nn\}$.
  In a certain interval of data all of the devices collect data $D = \{D1, D2, D3...,$
  $Dn\}$. The data is divided into n. Namely,

$$D_1 = \sum_{j=1}^{n} M_1 j,$$

$$D_2 = \sum_{j=1}^{n} M_2 j, \tag{7.5}$$

$$...,$$

$$D_3 = \sum_{j=1}^{n} M_n J,$$

7.4.1.2 Encryption and Exchange

- The job of each device is to keep one part and transmit the rest of the parts
  $\{Mi1, Mi2,..., Min(i \neq n)\}$ further. The secret keys, initially S1 and subsequently Sn, are used as following:

$$\begin{aligned} S_2 &= H\left(\tau \parallel S_1\right), \\ S_3 &= H\left(\tau \parallel S_2\right), \\ &..., \\ S_n &= H\left(\tau \parallel Sn - 1\right) \end{aligned} \tag{7.6}$$

- It is assumed that by using the key, one device Dv transmits the remaining part of the data Mij. A hash operation is done on Mij and ch = H(Mij ‖T). Mij and ch are then encrypted with Ki denoted as c = EKi (Mij ‖T‖ch). Then, Dv transfers the cipher text forwards.

7.4.1.3 Decryption and Confusion

- To get Mij back, cipher text c is decrypted with Si denoted as p = DSi (c). By doing a hash operation on Mij and T and matching the result with ch, the integrity of data received by the device can be verified. Wij will be discarded if it is not verified.
- The data D` of each gadget is the aggregate of its preserved slice Dii and received

n − 1 slices {D1i, D2i,..., Dni(i ≠n)} from other devices. The confused usage data of n devices is shown as following:

$$D_1' = M_{11} + \sum_{i=1}^{n} Mi_1$$

$$D_2' = M_{22} + \sum_{i=1}^{n} Mi_2 \qquad\qquad (7.7)$$

$$\dots$$

$$D_n' = M_{nn} + \sum_{i=1}^{n} min$$

7.4.1.4 Encryption and Reporting

- In the next step, the actual data is blinded and encrypted using Pk of the server. All blinded data D'_I, where i belongs from one to n, is encrypted, which is denoted as $Ci = E Pk (D'i)$, where i belongs from one to n.

- In the next step, a hash value $h_i = H_i(ID_i‖T_i‖C_i)$ is calculated, where H_i is a hash function, ID_i is the user's unique identification, T_i is the time, Ci is the preceding cipher text.
- This helps in resisting replay, impersonation, and manipulation attacks. In the final step, the device reports cipher text Ci and h to the aggregator.

7.4.1.5 Verification and Aggregation

- On obtaining cipher text Ci and the h_i, the aggregator confirms two things. First, if it is sent from a valid user within the network, (i.e. data repudiation), it checks the data's integrity. The aggregator performs a hash operation and if successful it accepts the cipher text Ci.

- The final result is $C_i = \prod_{i=1}^{n} Ci$. and it transmits that to the server. The server decrypts it with the private key S_k and obtains the total data $T = Dsk(C)$. This is the total data of all the IoT devices considered in a group in an area.
- Hence, the data aggregation is done successfully, where neither the server nor the aggregator is informed of other clients' data.

7.4.1.6 Security Analysis

Here, our methodology is scrutinized in the context of different attacks and how is privacy preserved in each scenario.

- **Eavesdropping Attack**
  An attacker can obtain the cipher texts of the data, but that is useless as the attacker does not have the private key. Without the private key, the cipher text cannot be decrypted. A brute force attack does not work in this scenario as symmetrical or asymmetrical encryption is applied to the data.

- **Replay Attack**
  Every message (M) has a hash value associated with it. This value is time dependent, say on T. If the message is tapped by an attacker A, it would not be verified when the aggregator or the device receives it. Hence, the aggregator will find out that a replay attack has been performed on the message.

- **Manipulation Attack**
  If any manipulation is done by an attacker on the message sent to the aggregator, it can be detected. Assume A sends a message M to another device B. The cipher text for it will be E(Mi||H(Mi||T). B decrypts the cipher text and obtains the hash value H(Mi||T) and transmitted data Mi. B then performs a hash operation on Mi and time T. this if matched with H(Mi||T), it means that the verification is successful and M is integral.

- **Impersonation Attack**
  In the event that a malicious user tries to impersonate a legitimate device sending data to the aggregator, then it can be distinguished.
  The adversary will need a valid ID for the device it is trying to impersonate, for example, device B. However, the ID of B is only known by B and the aggregator. Hence, the aggregator will verify the ID, and since the adversary will not have the ID, it will be detected.

- **Internal Attack**
  In this attack, the adversary is an internal curious device, such as an aggregator. It wants to obtain sensitive information from the IoT device. Let the

reporting message from all the devices be M={M1, M2…Mn}. M is encrypted with the public key of the server. Since no other internal device has access to this key, the cipher text cannot be decrypted. Hence, it is safe from internal attacks.

- **Colluding Attack**

  The above described scheme can resist collusion. This occurs among the many customers competing for resources. If the collected data of n users is M= {M1,M2,..,Mn} then the data is divided into n parts. A single user can only keep one part of the data; the remaining n–1 part is forwarded.

  The colluding users only know that Wij($i{\neq}j$) has been shared with them. No information is known of Wii, the one message that A holds. The Mi` is also not known to the colluding users, as the message is encrypted before sending it to the aggregator.

  Hence, even after having a number of colluding users, they cannot reveal another user's sensitive data.

7.4.1.7   Performance Evaluation

A single group consists of an aggregator connected with a vast number of users U = {U1,U2,...,Un}. While calculating the cost of operations, only the cost associated with encryption and decryption is studied. The cost of hash operations and addition operations is ignored as it is much lower.

The symbols below have the following meanings:

Ce = Computational cost of exponential operation
Cm = Computational cost of a multiplication operation
Co = computational cost of modulo operation
Ae = AES cost of encryption
Ad = AES cost of Decryption

For the data division, authentication, and aggregation cases no computation is needed when we are only dealing with a single device. Thus, the cost involves just the encryption and exchange phase operations, thus the cost is (n–1) Ae. In the next phase of decryption and encryption, a total of n–1 AES decryptions takes place, which makes the cost (n–1) Ad. The addition operations are ignored and not included while calculating the cost. The next phase is the encryption and reporting phase and the associated cost is 2Ce +Cm +Co. For the aggregator, the only cost involved is in the authentication and aggregation phase, which is (n–1)Cm.

In Figure 7.4, the time vs number of devices in an IoT network are plotted. The experiment is run in Python. Its minimum requirement is a 2.6GHz processor and 8 GB memory.

The decryption operation costs 2 ms. So the total computational cost can be denoted as S1 = n(n – 1) × 1 + n(n – 1) × 1 + n × 680, which is shown in Figure 7.4.

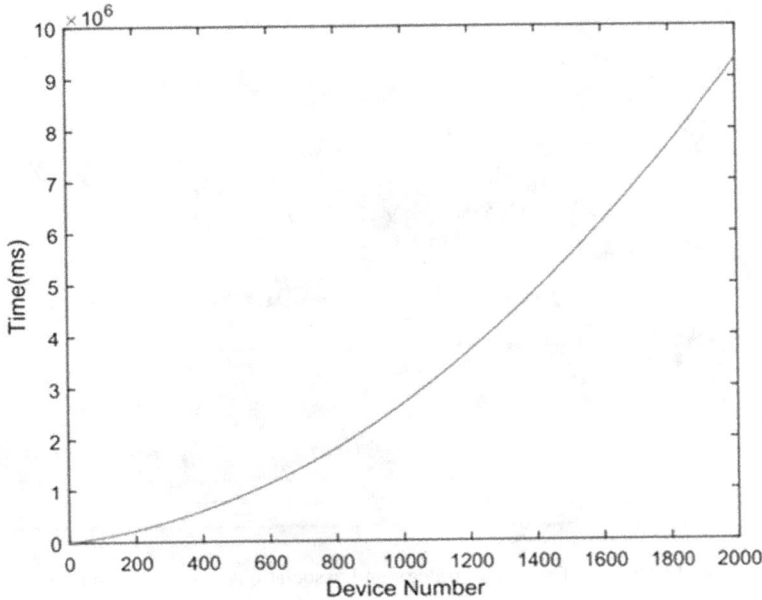

FIGURE 7.4 Total communication cost for the scheme.

7.4.2 EVOLUTIONARY GAME-BASED SECURE PRIVATE DATA AGGREGATION

The IoT has always been user-centric. This has become possible as it is heavily dependent on smartphones and vehicle sensors. They are part of mobile sensing computing devices and are showing the fastest growth as edge and connecting devices for the Internet and the IoT. In today's society, they are highly important as they work as the funnel for input sensor data [44].

One of the most serious threats to users' privacy is the data present in the mobile cloud sensing system. It has very sensitive information which can very easily spread via data channels and on social media application platforms. Hence, it is one serious privacy strain. In response, service providers (SPs) have to provide secure private data aggregation schemes. The SPs come up with plans to motivate their users to subscribe to these secure services. A substantial amount of work has been carried out at the ground level to promote secure behaviour over the system. This includes research focused on the cooperation between users [45].

Since the community-structured evolutionary game (CSEG) model is an evolutionary model, the security strategy evolves and spreads over the system. The users' behaviour is considered as strategies, which can be a secure strategy or a non-secure one. The users act as players so that they can spread their strategies. As shown in Figure 7.5, there is a community-structured topology.

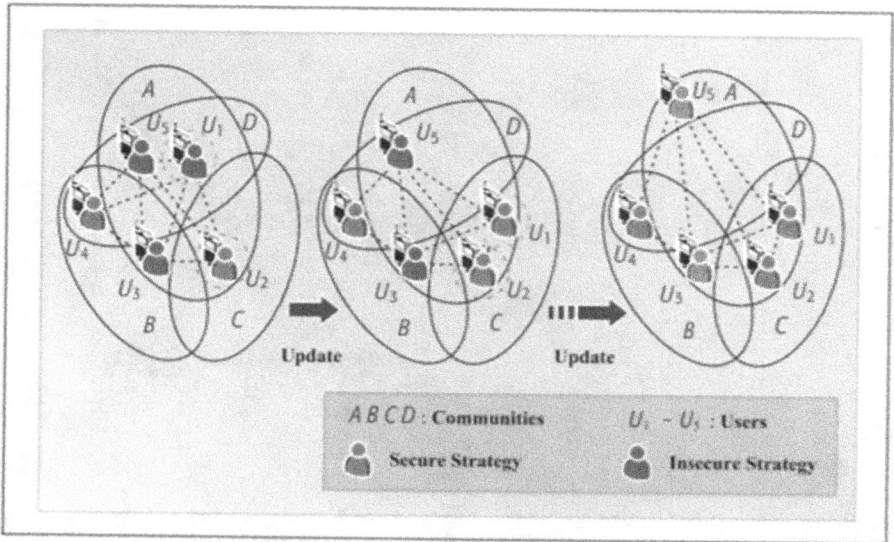

FIGURE 7.5 Evolution of security strategy and associations over a community-structured system.

Hence, the behaviour of one user in a community influences other users' choices in cases of utility and cost. A user can be part of more than one community. The user is considered to be rational, hence it sees other users in the community and chooses a strategy with high utility and the best privacy preserving mechanism. The critical cost performance (CCP) is then calculated by the SP for the privacy protection services.

7.4.3 SLICE-MIX-AGGREGATE AND IPDA

This scheme was first applied for security and data preservation purposes on wireless sensor networks [46]. It is based on a data slicing scheme, which means that collected data is sliced into blocks and then transmitted on different channels. This helps protect the sensitive data's integrity. Slice-mix-aggregate is also known as SMART. In this method of data slicing based data aggregation, the encrypted data is divided into pieces. These are sent to intermediate aggregate nodes. In the next step, they are combined with aggregate values. Similar steps are taken until the message is sent to the sink. Figure 7.6 describes the SMART scheme's core idea. Here, network size (N) = 5 (Nodes from S1 to S5), slicing size (J) = 2.

In the figure, **Dii** = piece of data kept by si,
$\qquad$ **Dij** = **transmitted bit** from si to sj,
$\qquad$ **ri** = data aggregated by si.

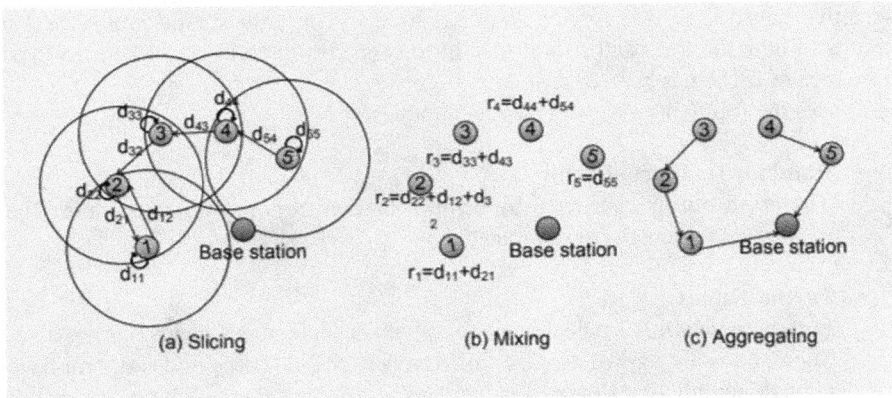

$r_4=d_{44}+d_{54}$

$r_3=d_{33}+d_{43}$

$r_5=d_{55}$

$r_2=d_{22}+d_{12}+d_3$

$r_1=d_{11}+d_{21}$

(a) Slicing   (b) Mixing   (c) Aggregating

FIGURE 7.6 Core idea of SMART.

There are a number of improvements that can be made to SMART scheme. For this, we have the integrity-protecting, private data aggregation (iPDA) scheme [47]. It improves the probity of information handled if instead only the SMART scheme was used. iPDA deals with the following shortcomings of SMART:

- High power consumption
- High chances of collision
- Collision loss

iPDA uses two disjointed aggregation trees (see Figure 7.7).It also uses slicing and assembly like SMART. Data aggregation is achieved by each node by sending

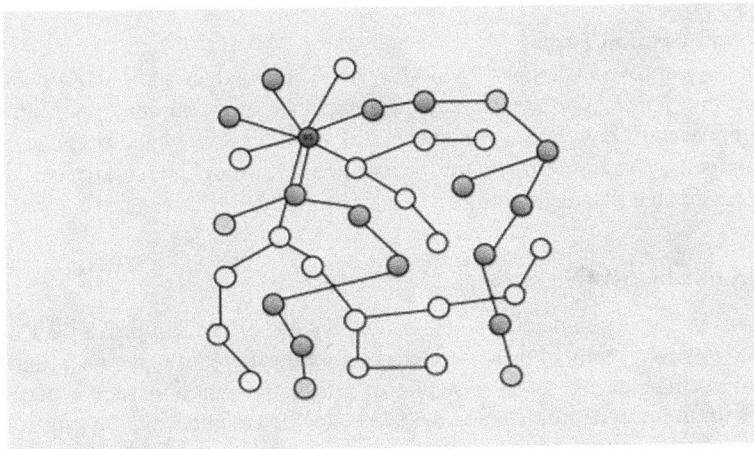

FIGURE 7.7 Two disjointed aggregation trees rooted at the base station.

sensitive data to the aggregation trees separately. The base station compares the results of both the trees and sees if they have been corrupted by any adversary. The core idea of iPDA has been described in Figure 7.7.

It uses the following compensation techniques:

- **Random Temporal Factor**
  The crash rate is decreased by using a random sending time schedule. The messages are never sent spontaneously.

- **Partial Factor**
  In total there are J–1 pieces to be sent, but for some nodes that is not possible. Such nodes are known as slice failed nodes (SFN). These nodes do not have enough neighbours. Hence, there will be some edges that will have to transmit a greater number of pieces. To ease up the process, the nodes are divided into two subsets, U and G, based on $x \geq J-1$, where x = 1, 2,...N. Set G consists of the SFN nodes. U consists of the remaining nodes.

- **Small Data Factor**
  In the iPDA scheme, a small piece of data is forwarded, in contrast to the SMART scheme, where J–1 pieces are forwarded to the neighbours. As only small pieces are sent out, the possibility of data loss on collision greatly reduces. A small data factor, L, is introduced. This defines the size of data to be sent to neighbours.

- **Sending Negative Factor**
  To increase the extent of data kept by the node, a negative piece is sent. It improves the aggregation accuracy as well. It additionally diminishes the impact of the information loss brought about by the crash.

- **Remuneration Factor**
  In the improved version of SMART, an acknowledgement (ACK) message is present. This is used to calculate the compensation factor and loss. Using ACK, a node can know about the successful transmission of the message. If not, compensation can be made for the loss in aggregation data. It also successfully forwards the data upstream.

7.5 CONCLUSION

Businesses and institutions constantly collect data to offer or improve their existing services. However, many of these services require the collection, analysis, and sometimes publishing/sharing of private sensitive data. Information privacy is particularly critical with ubiquitous information systems capable of gathering data from several sources, therefore raising privacy concerns with respect to the disclosure of such data. PPDM methods have been proposed to allow the extraction of knowledge from data while preserving the privacy of individuals. In this survey, an overview is provided of data mining methods that are applicable to PPDM of large quantities of information. This serves as preliminary background for the subsequent detailed description of the most common PPDM approaches. These PPDM methods are described according to the data life cycle phase at which they can occur, namely collection, publishing, distribution, and output of data. The evaluation of these techniques is then addressed by analysing metrics to assess the privacy and quality levels of data, as well as the complexity of the proposed PPDM approaches. Thereafter, the aforementioned PPDM techniques are considered from the point of view of their application to several practical domains and the rationale of their choice for those specific domains. Finally, some open issues and directions for future work are described.

Many businesses involved in domains such as smart healthcare [48], smart homes [49], and smart grids [50–51] collect users' data. This is done to give a better and personalized experience to the customers. To ensure that the data shared among these platforms is safe, we have discussed various methods. The privacy of users is of utmost concern when the web portal is collecting data from different sources, for example, in the case of data mining. Hence, the practice of privacy PPDM is taken up. PPDM methods have been discussed which enable big data to be obtained at a high volume and velocity. These methods included some recent and advanced methods like distributed privacy, one out of two oblivious transfer, and homomorphic encryption. Methods like homomorphic encryption and AES algorithms were discussed for secured and privacy preserving data aggregation, which lowers computational cost and improves the computation efficiency. Data aggregation using data slicing methods is also mentioned in depth [52]. Two methods, SMART and iPDA, and their advantages and disadvantages are discussed at length.

Finally, hybrid approaches such as evolutionary game-based secure private data aggregation guide us into future work. These applications help to maintain security and privacy when an IoT application interacts with non-IoT based wired or ad-hoc networks.

LIST OF ABBREVIATIONS

S.No	Abbreviation	Explanation
1	IoT	Internet of Things
2	DCL	Data collection
3	DPL	Data pre-processing layer
4	DML	Data mining output layer
5	Quasi-identifiers	QIDs
6	AES	Advanced encryption standard
7	SMART	Slice-mix-aggregate
8	iPDA	integrity-protecting, private data aggregation
9	KL	Kullback-Leibler
10	EMD	Earth mover's distance
11	CSEG model	Community-structured evolutionary game
12	CCP	Critical cost performance
13	SFN	Slice failed node
14	SP	Service provider

REFERENCES

[1] T. U. Darmstadt, *"Security and privacy challenges in industrial internet of things,"* in *Proceedings of 52nd Annual Design Automation Conference*, San Francisco, CA, USA, June 2015, 1–6.

[2] K. T. Nguyen, M. Laurent, and N. Oualha, "Survey on secure communication protocols for the Internet of Things," *Ad Hoc Netw.*, 32: 17–31, 2015.

[3] S. Sicari, A. Rizzardi, L. A. Grieco, and A. Coen-Porisini, "Security, privacy and trust in Internet of Things: the road ahead," *Comput. Netw.*, 76: 146–164, 2015.

[4] N. Zhang and W. Zhao, "Privacy-preserving data mining systems," *Computer*, 40: 52–58, Apr. 2007.

[5] F. K. Dankar and K. El Emam, "Practicing differential privacy in health care: A review," *Trans. Data Privacy*, 6(1): 35–67, 2013.

[6] C. Uhlerop, A. Slavković, and S. E. Fienberg, "Privacy-preserving data sharing for genome-wide association studies," *J. Privacy Confidentiality*, 5(1): 137–166, 2013.

[7] C. Lin, Z. Song, H. Song, Y. Zhou, Y. Wang, and G. Wu, "Differential privacy preserving in big data analytics for connected health," *J. Med. Syst.*, 40(4): 97, 2016.

[8] Z. Zhang, Z. Qin, L. Zhu, J. Weng, and K. Ren, "Cost-friendly differential privacy for smart meters: Exploiting the dual roles of the noise," *IEEE Trans. Smart Grid*, 8(2): 619–626, Mar. 2017.

[9] E. ElSalamouny and S. Gambs, "Differential privacy models for locationbased services," *Trans. Data Privacy*, 9(1): 15–48, 2016.

[10] P. Samarati and L. Sweeney, *"Protecting privacy when disclosing information: k-anonymity and its enforcement through generalization and suppression,"* in *Proceedings of IEEE Symposium on Research Security and Privacy*, Seattle,WA, USA, 1998, 384–393.

[11] P. Samarati and L. Sweeney, *"Generalizing data to provide anonymity when disclosing information,"* in *Proceedings of PODS*, New York, NY, USA, 1998, 188.

[12] M. M. Groat, W. Hey, and S. Forrest, *"KIPDA: k-indistinguishable privacy-preserving data aggregation in wireless sensor networks,"* in *Proceedings of IEEE INFOCOM*, Apr. 2011, Shanghai, China, 2024–2032.

[13] A. R. Beresford and F. Stajano, "Location privacy in pervasive computing," *IEEE Pervasive Comput.*, 2(1): 46–55, Jan./Mar. 2003.

[14] B. Bamba, L. Liu, P. Pesti, and T. Wang, *"Supporting anonymous location queries in mobile environments with privacygrid,"* in *Proceedings of the ACM 17th International Conference on World Wide Web*, Beijing, China, 2008, 237–246.

[15] X.-M. He, X. S. Wang, D. Li, and Y.-N. Hao, "Semi-homogenous generalization: Improving homogenous generalization for privacy preservation in cloud computing," *J. Comput. Sci. Technol.*, 31(6), 1124–1135, 2016.

[16] T. S. Gal, Z. Chen, and A. Gangopadhyay, "A privacy protection model for patient data with multiple sensitive attributes," *Int. J. Inf. Secur. Privacy*, 2(3): 28, 2008.

[17] B. C. M. Fung, K. Wang, R. Chen, and P. S. Yu, "Privacy-preserving data publishing: A survey of recent developments," *ACM Comput. Surveys.*, 42(4): 14:1–14:53, 2010.

[18] M. Xue, P. Kalnis, and H. K. Pung, "Location diversity: Enhanced privacy protection in location based services," in *Location and Context Awareness*. Berlin, Germany: Springer-Verlag, 2009, 70–87.

[19] F. Liu, K. A. Hua, and Y. Cai, *"Query l-diversity in location-based services,"* in *Proceedings of the IEEE 10th International Conference on Mobile Data Manage., System, Services Middleware (MDM)*, Taipei, Taiwan, May 2009, 436–442.

[20] Z. Zhang, Z. Qin, L. Zhu, J. Weng, and K. Ren, "Cost-friendly differential privacy for smart meters: Exploiting the dual roles of the noise," *IEEE Trans. Smart Grid*, 8(2), 619–626, Mar. 2017.

[21] E. ElSalamouny and S. Gambs, "Differential privacy models for locationbased services," *Trans. Data Privacy*, 9(1), 15–48, 2016.

[22] H. AbdulKader, E. ElAbd, and W. Ead, "Protecting online social networks profiles by hiding sensitive data attributes," *Procedia Comput. Sci.*, 82: 20–27, Mar. 2016

[23] L. Chang and I. S. Moskowitz, "*Parsimonious downgrading and decision trees applied to the inference problem*," in *Proceddings of the Workshop New Secur. Paradigms*, 1998, 82–89; R. Mendes, J. P. Vilela, "Privacy-Preserving Data Mining: Methods, Metrics, and Applications," 5, 10580, 2017.

[24] A. A. Hintoglu and Y. Saygın, "Suppressing microdata to prevent classification based inference," *VLDB J.-Int. J. Very Large Data Bases*, 19(3): 385–410, 2010.

[25] R. Agrawal and C. Johnson, "Securing electronic health records without impeding the flow of information," *Int. J. Med. Inform.*, 76(5–6): 471–479, 2007.

[26] S. Even, O. Goldreich, and A. Lempel, "A randomized protocol for signing contracts," *Commun. ACM*, 28(6): 637–647, 1985.

[27] M. Naor and B. Pinkas, "*Efficient oblivious transfer protocols*," in *Proceedings of the 12th Annu. ACM-SIAM Symp. Discrete Algorithms*, Philadelphia, PA, USA, 2001, 448–457.

[28] J. R. Troncoso-Pastoriza, S. Katzenbeisser, and M. Celik, "*Privacy preserving error resilient DNA searching through oblivious automata*," in *Proceddings of the 14th ACM Conference on Computer and Communication Security*, Alexandria, VA, USA, 2007, 519–528.

[29] J. Vincent, W. Pan, and G. Coatrieux, "*Privacy protection and security in ehealth cloud platform for medical image sharing*," in *Proceddings of the IEEE 2nd International Conference on Advanced Technology Signal Image Process. (ATSIP)*, Monastir, Tunisia, Mar. 2016, 93–96.

[30] M. S. Kiraz, Z. A. Genç, and S. Kardas, "Security and efficiency analysis of the Hamming distance computation protocol based on oblivious transfer," *Secur. Commun. Netw.*, 8(18), 4123–4135, 2015.

[31] G. Taban and V. D. Gligor, "*Privacy-preserving integrity-assured data aggregation in sensor networks*," in *Proceddings of the IEEE International Conference on Computer Science Engineering (CSE)*, vol. 3, Vancouver, BC, Canada, Aug. 2009, 168–175.

[32] N. Cao, C. Wang, M. Li, K. Ren, and W. Lou, "Privacy-preserving multikeyword ranked search over encrypted cloud data," *IEEE Trans. Parallel Distrib. Syst.*, 25(1), 222–233, Jan. 2014.

[33] R. Sheikh, B. Kumar, and D. K. Mishra. (2010). "A distributed k-secure sum protocol for secure multi-party computations," *International Journal of Computer Science and Information Security*, 6(2), 184–188, Nov. 2009.

[34] R. Sheikh and D. K. Mishra, "*Secure sum computation for insecure networks*," in *Proceedings of the 2nd International Conference on Information and Communication Technology Competitive Strategies*, Udaipur India, 2016, 102.

[35] T. Tassa, "Secure mining of association rules in horizontally distributed databases," *IEEE Trans. Knowl. Data Eng.*, 26(4): 970–983, Apr. 2014.

[37] S. Choi, G. Ghinita, and E. Bertino, "*A privacy-enhancing contentbased publish/subscribe system using scalar product preserving transformations*," in *Proceedings of the International Conference on Database Expert System Application*, Linz, Austria, 2010, 368–384.

[38] M. J. Freedman, K. Nissim, and B. Pinkas, "*Efficient private matching and set intersection*," in *Proceedings of the International Conference on Theory anf Application Cryptograph Technology*, Interlaken, Switzerland, 2004, 1–19.

[39] H. Rong, H.-M. Wang, J. Liu, and M. Xian, "Privacy-preserving k-nearest neighbor computation in multiple cloud environments," *IEEE Access*, 4: 9589–9603, 2016.

[40] Fontaine, C. and Galand, F., "A survey of homomorphic encryption for nonspecialists," *EURASIP J. Inf. Secur.* 2007(1): 1–10, 2007.

[41] Paillier, P., "Public-key cryptosystems based on composite degree residuosity classes," in: Stern, J. (ed.) *EUROCRYPT 1999. LNCS*, vol. 1592, 223–238. Heidelberg: Springer, 1999.

[42] X. Hei et al., "*PIPAC: Patient Infusion Pattern Based Access Control Scheme for Wireless Insulin Pump System*," *IEEE INFOCOM'13*, Turin, Italy, Apr. 14–19, 2013, 3030–3038.

[43] J. Du et al., "Community-structured evolutionary game for privacy protection in social networks," *IEEE Transactions on Information Forensic and Security*, vol. 13, Mar. 2018, 574–589

[44] H. E. Wenbo, L. Xue, H. Nguyen et al., "*Pda: Privacy-preserving data aggregation in wireless sensor networks*," in *Proceedings of the 26th IEEE International Conference on Computer Communications (IEEE INFOCOM'07)*, NW Washington, DC, USA, 2007, 2045–2053.

[45] D. Westhoff, J. Girao, and M. Acharya," Concealed data aggregation for reverse multicast traffic in sensor networks: Encryption, key distribution, and routing adaptation, *IEEE Trans. Mobile Comput.*, 5(10): 1417–1431, 2006.

[46] J. Xu, G. Yang, Z. Chen, and Q. Wang, "A survey on the privacy-preserving data aggregation in wireless sensor networks," *China Commun.*, 12: 162–180, 2015. 10.1109/CC.2015.7112038.

[47] J. Du, C. Jiang, E. Gelenbe, L. Xu, J. Li, and Y. Ren, "Distributed data privacy preservation in IoT applications," *IEEE Wireless Commun.*, 25: 68–76, 2018. 10.1109/MWC.2017.1800094.

[48] M. Salam, W. C. Yau, J. J. Chin, S. H. Heng, H. C. Ling, R. Phan, G. S. Poh, S. Y. Tan, and W. S. Yap, "Implementation of searchable symmetric encryption for privacy-preserving keyword search on cloud storage," *Human-centric Comput. Inform. Sci.*, 5: 19, 2015. 10.1186/s13673-015-0039-9.

[49]. A. Alrawais, A. Alhothaily, C. Hu, and X. Cheng, "Fog computing for the internet of things: security and privacy issues," *IEEE Internet Comput.*, 21(2): 34–42, 2017.

[50]. H. Bao and R. Lu, "A new differentially private data aggregation with fault tolerance for smart grid communications," *IEEE Internet Things J.*, 2(3): 248–258, 2015.

[51] C. Hu, H. Liu, L. Ma, Y. Huo, A. Alrawais, X. Li, H. Li, and Q. Xiong, "A secure and scalable data communication scheme in smart grids," *Wirel. Commun. Mob. Comput.*, 2018: 1–17, 2018.

[52] T. Song, R. Li., B. Mei, J. Yu, X. Xing, and X. Cheng, "A privacy preserving communication protocol for IoT applications in smart homes," *IEEE Internet Things J.*, 4(6): 1844–1852, 2017.

8 Real-Time Cardiovascular Health Monitoring System Using IoT and Machine Learning Algorithms
Survey

T. Vairam, S. Sarathambekai, K. Umamaheswari, R. Jothibanu, and D. Manojkumar
PSG College of Technology, India

8.1 INTRODUCTION

In general, the Internet of Things (IoT) in healthcare systems has benefits such as: i) the workload of doctors, nurses, or medical practitioners is reduced, ii) it makes healthcare more reasonable, iii) it makes doctors more approachable and accessible for rural patients, and iv) it encourages progress in medical technology. According to the World Health Organization (WHO) [1], out of all diseases, cardiovascular disease (CVD) is the main cause of death among humans globally. The 2019 annual report about heart disease [2] says that CVD is one of the major causes of death in the US. Out of nearly 840,768 deaths in 2016, 635,260 demises were due to CVD. Though the death rate has decreased nowadays, still it requires more attention. Nowadays, patients' health is being monitored remotely using IoT technologies and devices. This enables the doctors to take care of patients all the time without actually being physically present. The communication between the doctor and patient becomes easier and more efficient than with the traditional system where the patient needs to meet the doctor in person often. This is a time-consuming process. The era of the IoT has made the healthcare system smarter. Figure 8.1 shows the general process involved in the IoT healthcare system.

The objectives of this chapter are to analyse how efficiently IoT can play a role in healthcare systems, especially for CVD, to review the machine learning (ML) algorithms used in this field, and to provide the abstraction of edge/fog computing in the IoT healthcare field. Section 8.2 brief discusses CVD and its types. Section 8.3 describes the motivation and classification. Section 8.4 contains the comparison of

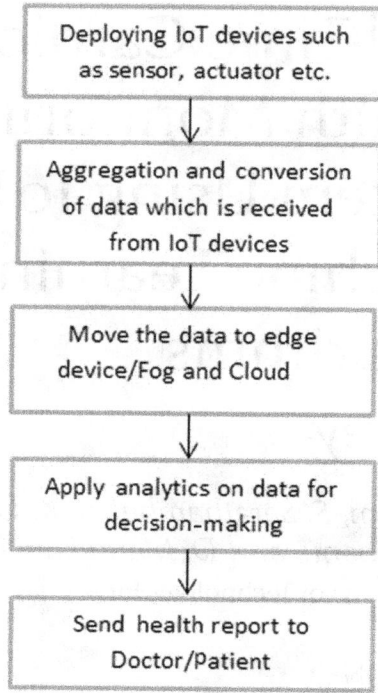

FIGURE 8.1   General process in the IoT healthcare system.

healthcare monitoring systems under the IoT. Section 8.5 discusses the ML algorithms implemented in CVD healthcare monitoring systems and Sections 8.6 and 8.7 deal with the role of edge/fog computing in IoT healthcare systems and issues and challenges in IoT healthcare systems, respectively.

8.2 CARDIOVASCULAR DISEASES (CVD)

A collection of diseases affecting the heart and blood vessels can be defined as CVD. Types of CVD are:

- Rheumatic Heart Disease: This form of CVD permanently damages the heart because the heart malfunctions and damages the heart valves.
- Congenital Heart Disease: This is another type of CVD which generally affects the new-borns by means of a deformity of the heart. The different types of this heart disease are :
  - Atrial Septal Defect (ASD)
  - Ventricular Septal Defect (VSD)

The most widespread kind of heart syndrome is coronary heart syndrome [1]. Over 54.5 million human lives were affected by this form of CVD in 2016.

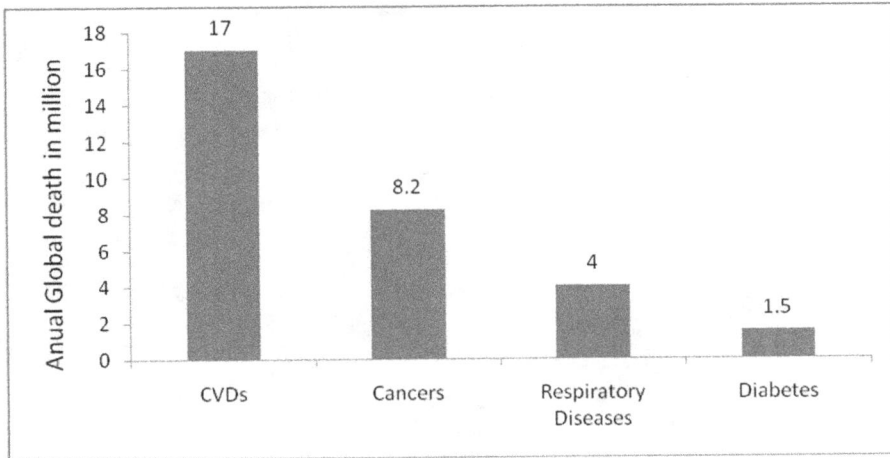

FIGURE 8.2   Death rate.

According to the data, CVD is the most common cause of death [3]. Figure 8.2 shows the death rate of CVD compared with other deadly diseases such as cancer. This shows the potential impact of implementing the healthcare monitoring for CVD.

8.3   MOTIVATION AND CLASSIFICATION

The main objective of this chapter is to analyse the various methodologies to prevent and monitor details of CVD-affected patients in real-time with the IoT. Patients face unexpected death from the specific cause of heart problems and heart attacks, due to lack of adequate medical care for patients at the required time. Hence, the system is required to monitor the patient continuously and predict the CVD. The IoT model is used to monitor the patient's heart rate and temperature using a pulse and DTH11 sensor. The ML algorithm is used for predictive analysis.

8.3.1   INTERNET OF THINGS (IoT)

Every object in a network is provided with a unique identification. An object can be a digital device, animal, human, and so on. When these are interconnected and transfer the information through the Internet, it is referred to as the IoT. M2M stands for Machine-to-machine communication, namely a machine connected with another machine by itself that does not require a human to be involved for further processing. M2M means connecting, managing, and collecting data from the cloud. M2M provides the IoT-enabled connectivity, which helps people to work smarter and to gain full control over their day-to-day activities. Devices and artefacts are linked to a network for the IoT, and its responsibility is to amalgamate information from the diverse devices and to share the meaningful information according to the application needs. The meaningful information is obtained by applying an analytics-based algorithm in the data that is received from the IoT devices.

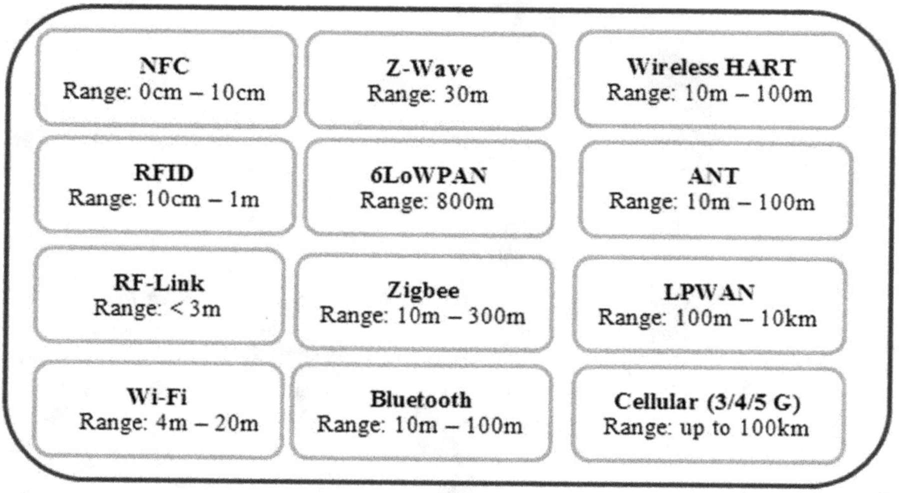

FIGURE 8.3 Communication technologies and their ranges.

Evolution of the IoT and the impact of IoT in the agriculture field have been explained by Khanna and Karur [4]. According to the data given by Rehman et al. [5] and in *Communication Technologies* [6], Figure 8.3 represents the various communication technologies and their communication ranges used in the IoT applications. The technologies given in Figure 8.3 are not limited. There are more technologies available in the IoT field. The various communication technologies and their applications have been addressed by various researchers [7–16]. Some of the applications are listed in Table 8.1.

8.3.2 IoT Applications

The IoT promises to bring great value to our lives. The existence of new-fangled wireless networks, better-quality sensors, and innovative computing capability means the IoT could be the next cutting edge in the race for your wallet. Billions of every-day items are provided with connectivity and intelligence. It is already widely used in several fields. Various smart applications using IoT have been discussed by researchers [17–22]. Some of the IoT applications are given in Figure 8.4.

8.3.3 IoT in Healthcare

Supported devices on the IoT assist with remote monitoring in the medical system. Their potential to keep patients in safe hands and well and allow doctors to give better care of patients is very significant. IoT in healthcare allows patients to experience greater commitment and satisfaction with doctors as it is becoming easier for the patient to approach the doctor. IoT in healthcare has a major impact on money, as it helps to reduce costs, which leads to better treatment being offered.

The IoT plays a major role in healthcare applications that support sick people and their family members, medical doctors, and hospitals. Figure 8.5 represents the

TABLE 8.1

Communication technologies and their applications

Communication Technologies	Applications
NFC (Near Field Communication)	• Digital out of home (DOOH) • Electronic visitor verification (EVV)
Z Wave	• Smart hubs • Smart lighting • Smart locks
wireless HART	• Smart industry
RFID	• People tracking • Document tracking • Healthcare • Smart libraries
6LoWPAN	• General automation • Smart grids • Home automation • Industrial monitoring
ANT	• Multicast wireless sensor network
RF Link	• Wireless data • Voice transfer applications
Zigbee	• Smart asset tracking • Smart grid monitoring
LPWAN	• Smart agriculture • Smart building applications
Wi-Fi	• Hotspots • Industries – inventories, positioning
Bluetooth	• Audio signal transmission • Smart homes • Wearable devices • Barrier gates

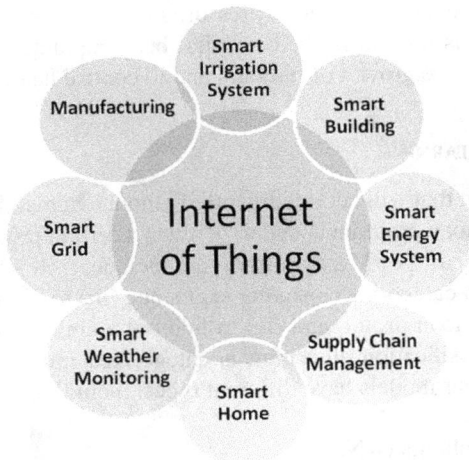

FIGURE 8.4 IoT applications.

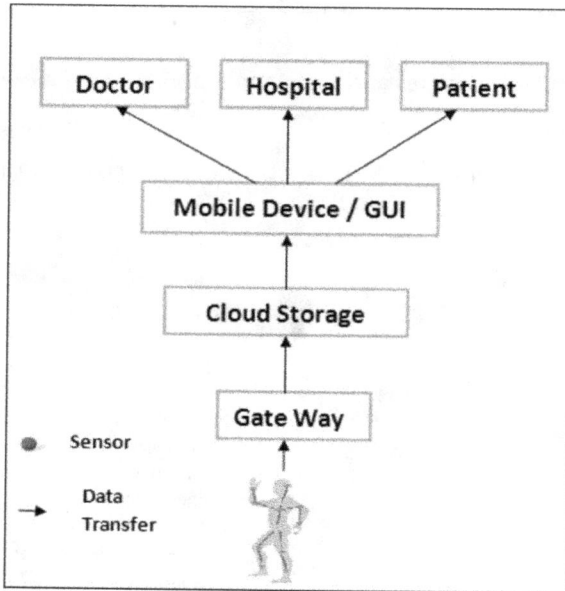

FIGURE 8.5   IoT in healthcare.

process involved in a healthcare system using IoT. Using IoT in healthcare, the patient can communicate with doctors in a trouble-free manner. Sensors measuring aspects such as ECG, temperature, respiratory rate, blood pressure, pulse, and blood sugar levels, play a major role in smart healthcare monitoring systems. Figure 8.6 shows images of sensors used in a healthcare system. In the patient monitoring application, the blood pressure and blood sugar sensors are used. The sensor are used to monitor heart and brain activity. The biosensor is used to observe biological components of the human such as blood sugar and cholesterol levels, and for pregnancy testing. Pressure sensors used for blood pressure monitoring systems are also used to detect that a patient is present in their hospital bed. Implanted sensors are used to monitor and measure the growth of tumours and to control heart pacemakers.

8.3.4 MACHINE LEARNING

The computer system that will act intelligently without a human involved in its functionality and that is not a program coded to perform a certain process is referred to as ML. The various use cases of ML algorithms are include: (i) to identify that an email contains spam, (ii) to categorize consumer segments, (iii) to discover whether a person got her/his bank loan or not, and (iv) to find out a child's exam results [23]. In general, the ML classification algorithm is categorized into linear and non-linear models. The non-linear models have different types, including:

- K-nearest neighbours (KNN)
- Naïve Bayes
- Random forest classification.

Temperature Sensor	Pulse Sensor	Blood Pressure Sensor
Blood Sugar Sensor	ECG	Respiratory Rate

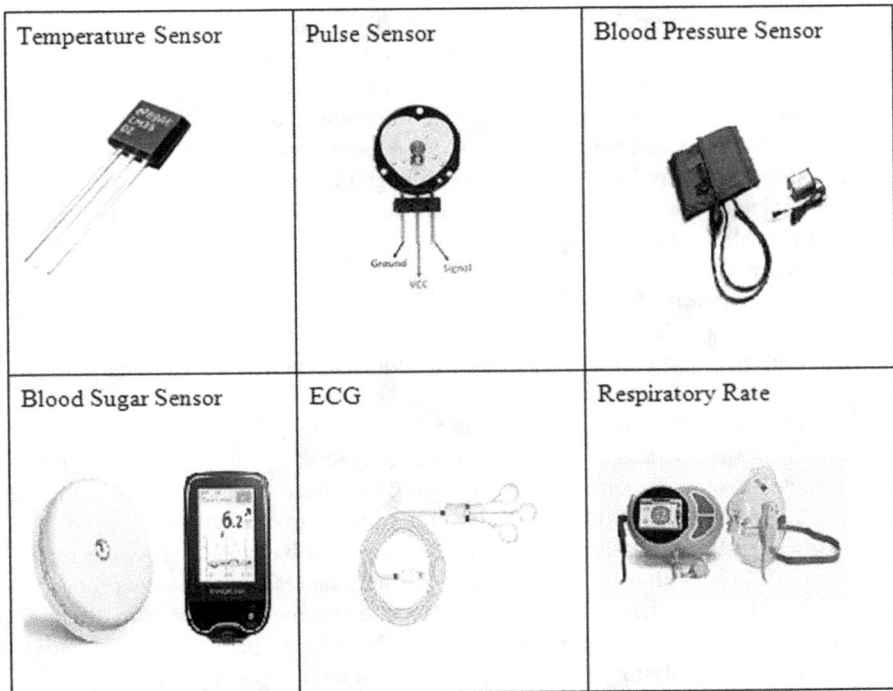

FIGURE 8.6 Images of sensors in healthcare system.

The role of these algorithms is to specify which class the data belongs to. In other words, they are used to predict the class for an input data.

8.3.5 MACHINE LEARNING IN HEALTHCARE

Healthcare has to be considered critical work, providing value-based care to millions of people across the world. An innovative idea is required for the healthcare specialists and stakeholders to deliver high-quality services. Nowadays, Internet-connected medical devices are used in the smart healthcare system using the latest technology. ML in healthcare is one of the recent trends which are gradually being accepted by the healthcare industry. Researchers have identified the ten healthcare applications in which ML plays a crucial role [24–26]. These are:

Identifying Diseases and Diagnosis [27]: Still there is a problem with diagnosis which leads to 1:10 patient dying in hospital. ML plays a most important role in the healthcare system and is being used to identify rare diseases such as pneumonia, cancer, and other diseases.

Medical Imaging Diagnosis [28]: ML algorithms can also be used for computing the medical image features which are to be considered in predicting and diagnosing the diseases. The ML algorithm starts with identifying the image feature for classification and then performs computation on the classified feature based on metrics according to the disease.

Drug Discovery and Manufacturing [29]: Drug discovery and development is an intricate process which depends on several factors in the healthcare system. Efficient computations in drug discovery lead to reductions in the cost of bringing new drugs to the market. ML techniques offer tools that help find a better way of discovering drugs and verdicts, making for queries with plentiful data.

Precision Medicine [30]: By referring to huge multidimensional biological data sets, clinicians can specify the procedure for diagnosis. This is called precision medicine. ML algorithms support clinicians to cautiously adapt early intervention with patients.

Machine Learning-Based Behavioural Modification [31]: Behaviour analysis means reviewing the deportment or behaviour of a person based on the profile of a particular person built earlier. The ML algorithm helps by retrieving the data efficiently.

Smart Health Records [32, 33]: There will always be a high demand for enlightening healthcare services. Smart health records are one of the important applications and are a core component in improving healthcare services. ML algorithms are used to aid smart health records to offer various services efficiently.

Clinical Trial and Research [34]: Clinical trials involve conducting experiments in a clinical research centre. ML algorithms help the researcher easily predict the outcome of clinical trials, which in turn helps to approve the novel drug on time.

Crowd Sourced Data Collection [35, 36]: This kind of data collection involves many people in solving problems. This helps in examining a particular problem globally. The role of ML is to perform effective data classification from the large set of data.

Better Radiotherapy [37, 38]: ML has the potential to transform the field of radiation oncology. Object classification can be done in an efficient and easy way using an ML algorithm. The process of diagnosis and finding variables from multiple samples can also be done easily using ML algorithms.

Outbreak Prediction [39]: Examining and forecasting pandemics or epidemics across the world is important in all kinds of situation. ML algorithms help to collect all kinds of information pertaining to pandemic diseases and predict all infectious diseases.

8.4 COMPARISON OF HEALTHCARE MONITORING SYSTEMS UNDER IoT

Much research has been undertaken in the field of IoT in healthcare. This section identifies some of the work pertaining to the IoT in healthcare. The objective of Verma and Sood [40] is to provide a monitoring system for the remote patient in the smart home environment, using a classification algorithm at fog level and temporal mining at cloud level for event triggering activity, and finally to provide an alert system for the doctors and caregivers using a decision-based approach. The overall benefits of their proposed system are to reduce the time delays so that they respond quickly to the patient's monitored data [41]. They predict the hypertension attack according to the patient's symptoms. The symptoms are collected from the patient's body using various sensors. For implementation, three different data sets are used. They provide the fog computing facilities for sending the alert to the doctors.

A smart health system is proposed by Rani et al. [42] to identify the chikungunya infections. Their system contains a mobile app to view the precaution measures, sensors, and objects (people and devices) to collect the data, and uses communication technologies for efficient data transaction. Ding et al. [43] concentrated on the security aspects of the patient health data. They developed a secure system for health data storage and use data integrity verification methods to ensure that the system is highly secure. The main contribution of Yang et al. [44] is i) a wearable biosensing mask developed to capture the facial expressions of the patients who are hospitalized, (ii) a mobile app developed for the purpose of recording, processing, interpreting and viewing the data in an understandable format, and (iii) evaluation of the pain states using multi-channel facial EMG. An efficient system for heart rate variability is proposed by Hussein et al. [45]. The data from the ECG device is considered. The research work by Vadrevu and Manikandan [46] contributed a secure Edge of Things framework for the healthcare system. A fully homomorphic encryption technique is used to provide the assurance of security. Figure 8.7 shows the comparison of healthcare systems using IoT between various research works.

8.5 MACHINE LEARNING ALGORITHMS IMPLEMENTED IN CVD HEALTHCARE MONITORING SYSTEM

ML is making the computer systems which can be used in the same way to make data relevant to humans. ML is a form of artificial intelligence, in simple terms, that uses an algorithm or method to extract patterns of raw data. The main objective of ML is to permit computers to be trained from practices without precise programming or human involvement. Various research into healthcare monitoring systems has been undertaken [47–52]. A patient monitoring system is suggested by Ani et al. [53] for patients with strokes to prevent potential recurrence by alerting the consultant and concierge by detecting differences in risk factors for stroke diseases. This system uses the diagnosis and prediction classification algorithm. Random forest's tree-based classification system provides 93% accuracy. A microcontroller linked to several portable sensors and the cloud implements the device. Entries are collected using sensor devices and are transferred to cloud storage, which generates the warning message. They used different classification algorithms, including naïve Bayes, KNN, and decision trees. The result shows that the algorithm for random forest is more reliable than the algorithms for naïve Bayes, KNN, and decision trees.

Atrial fibrillation [47] is found through the monitoring of heartrate fluctuations. Light-based sensing is used to detect the non-invasive Photoplethysmogram (PPG) sensor heart rate fluctuations. The observed signal is processed further to remove noise in the signal. The framework also offers IoT infrastructure for the monitoring of atrial fibrillation using the IoT application programming interface API. The software (API) used for cloud analysis is the height of things. Tan et al. [48] proposed a hybrid approach that effectively combines the wrapping approaches of the support vector machine (SVM) and genetic algorithm. The findings of this approach are analysed using the data mining methods Library for Support Vector Machines (LIBSVM) and Weka. Otoom et al. [49] proposed a device that detects and tracks coronary artery disease. Parthiban and Srivatsa [50] used the ML algorithms in this program for diagnosis of heart disease in diabetes patients. All of these methods

Paper	Machine Learning algorithms	Disease	Real Time Data	Data set	Fog/Cloud/Edge	Implication	Security
[16]	Bayesian Belief Network classifier	Fever, BP, Glucose level	Yes	No	Fog, Cloud	Reduced time delay	No
[17]	Artificial Neural Network	BP fluctuation, Hypertension Attack	No	Yes	Fog, Cloud	Alert is being sent to the doctor quickly	No
[18]	Fuzzy K-Nearest Neighbour	Chickungunya Virus	Yes	No	Edge, Cloud	More number of patients can be monitored at the location basis. Quick recognition of chickungunya infections	Yes
[19]	-	Patient Health data	Nil	Nil	Edge	Computation cost of the system is reduced.	Yes
[20]	-	Pain intensity	Yes	Nil	Cloud	The patients pain states are evaluated using multi-channel facial sEMG	No
[21]	-	Heart Rate Variability(H RV)	No	Yes	cloud	HRV analysis has done efficiently. Security features also added	Yes
[22]	-	Photoplethys mogram (PPG) signal	Yes	Yes	-	Achieved a false alarm rate reduction(FARR) with	No
[23]	Clustering based Techniques	Patient health data	No	Yes	Edge, Cloud	Edge of Things framework is developed for healthcare reconnaissance services	Yes

FIGURE 8.7   Comparison of IoT healthcare systems.

have the common goal of classifying cardiovascular disease using classification techniques. The suggested solutions provide a way to obtain an efficient accuracy level by taking algorithms such as KNN and the random forest algorithm with specific ML.

8.5.1 IMPLEMENTATION OF RANDOM FOREST AND KNN FOR CVD HEALTH DATA

The proposed methodology is shown in Figure 8.8. The process begins from observing the health data of the human using a pulse sensor and DTH11 sensor. The sensor value is sent through NodeMCU to a mobile device and it will be made available in the cloud storage network. For prediction of disease, the ML algorithms such as random forest and KNN are used. After predicting the disease, the report is sent to the doctor as well as to the patient. The data from the data set is obtained at the initial stage. Pre-processing methods are applied at the initial stage to identify whether the person is suffering from diseases or not, then feature extraction and selection techniques are applied to detect CVD. Eventually, the classifiers random forest and KNN were improved to assess the heart disease rates. Such classifications come from the data-related alternative features but are much shorter than the features that exist. The sensed data can be processed in the cloud system and stored there. ML algorithms can be used to predict the data. The data set is taken from the archive in learning about heart disease on the UCI Heart database. There are 76 attributes and 303 records in the data set. For this analysis and research, only 14 attributes are used. The characteristics of the attributes are actual, full, and categorical.

8.5.2 IMPLEMENTATION RESULTS

In this implementation, trial and error has been performed based on the database. This proposed method is validated on UCI Heart databases; the consummation of the mechanism was calculated based on accuracy value, sensitivity, and specificity.

$$Sensitivity = \frac{TP}{TP + FN} \qquad (8.1)$$

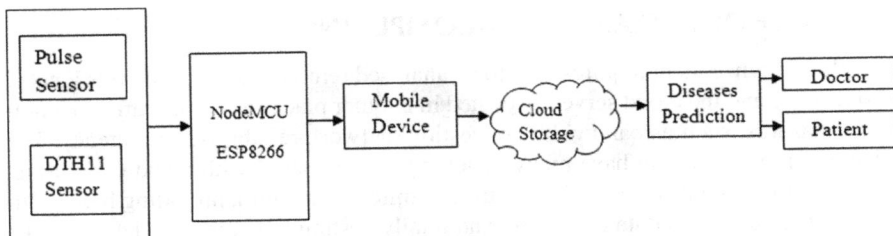

FIGURE 8.8 CVD smart healthcare system.

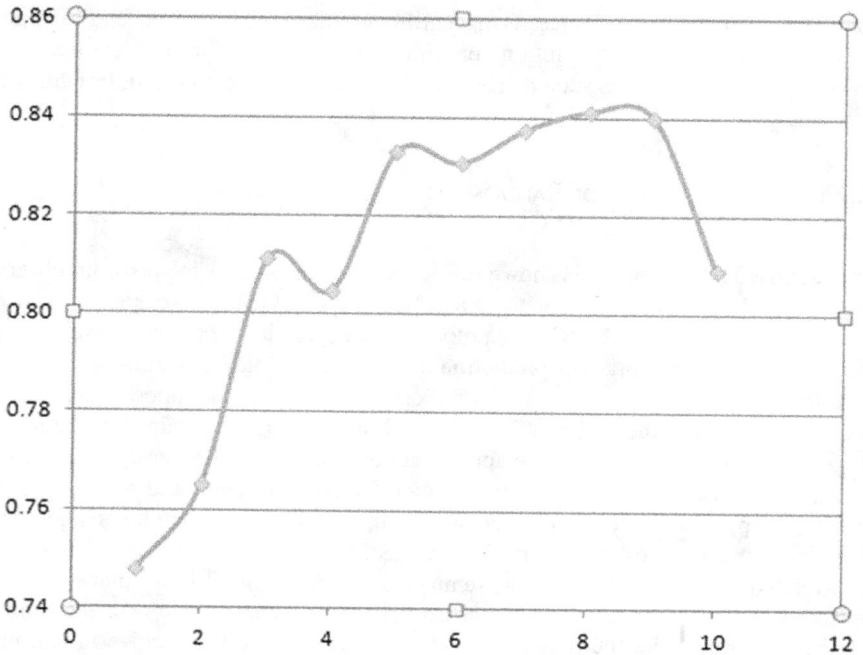

FIGURE 8.9   Accuracy level of KNN.

$$Specificity = \frac{TN}{TN + FP} \tag{8.2}$$

$$Accuracy = \frac{TN + TP}{TN + FP + TP + FN} \tag{8.3}$$

Where TP = True Positive, TN = True Negative, FN = Fake Negative, FP = Fake Positive

Figure 8.9 shows the accuracy level of KNN. The accuracy levels for the random forest algorithm and KNN algorithm are 88% and 79%, respectively. Based on that, the random forest algorithm gives a greater accuracy level than the KNN.

8.6   ROLE OF FOG AND EDGE COMPUTING

The cloud environment requires data to be analysed remotely as the patient is located in one place and the cloud server is located in another place [54]. This carries a number of hazards, such as bandwidth congestion, network reliability, and latency. The edge and fog computing have many benefits [55] compared with cloud computing, such as achieving fast retransmission, using limited bandwidth, providing high security, having control on data generated, and finally costing less. Figure 8.10 shows the architecture for the edge-based healthcare system proposed by Oueida et al. [56].

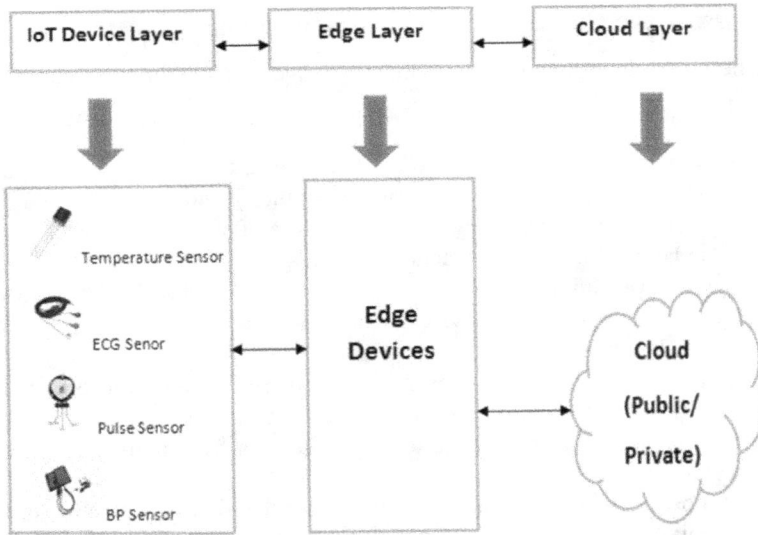

FIGURE 8.10   Architecture for edge-based healthcare system.

Researchers have stated that the cloud and edge computing are major reasons for successful deployment of smart healthcare systems. The mathematical model for the healthcare system using a petri net is also given by Soraia Oueida et al. [56]. The research work by Mahulea et al. [57] presented a modular approach for modelling healthcare systems using a petri net and proved that the petri net modelling technique can also be used for constructing healthcare systems. Chen et al. [58] proposed the edge cognitive computing-based smart healthcare system. Resource allocation for the edge devices is one of the important considerations when connecting more sensors and transferring their data.

8.7   ISSUES AND CHALLENGES

8.7.1   General Issues in Machine Learning Algorithms

ML is recommended for any application where it will provide the necessary level of accuracy for the business. However, ML also has some issues as follows:

- the usage of low-quality data causes problems with pre-processing data and data extraction functionality;
- the functions of recording, extracting, and restoring data are time-consuming;
- the algorithms become archaic: as the data size for many applications increases rapidly, the algorithms may not be able to handle the new data sets; and
- some data generated by the application may not be relevant or valuable, due to poor data conversion. If the ML algorithms fail to understand the data properly, the algorithms will provide unexpected results.

8.7.2 Issues and Challenges of IoT in Healthcare

Researchers have identified the following challenges for IoT in healthcare applications [59]:

- Data security and privacy: IoT devices are meant for capturing data in real-time, as most IoT devices concentrate on capturing and transferring data to the next level, but their design lags behind in terms of efficient data transfer protocol standards which provide security features;
- Integration of multiple devices: the IoT in healthcare involves more than one device, and integration of devices is an encumbrance in the realization of IoT in the healthcare system;
- Integration of protocols: not every IoT device uses the same protocol standards, and this unevenness leads to a slowdown in the process;
- Data aggregation: it is intricate to aggregate data from non-uniform data for analysis; and
- Cost: cost is one of the challenges to be considered and plays a major role in developing IoT healthcare systems.

8.7.3 Issues and Challenges in Fog and Edge Computing

More research challenges are posed by fog/edge computing [60–64]. Al-Khafajiy [65] has identified the challenges of fog computing as:

- resource management: this is one of the greatest challenges in fog computing, because the computing and storage resources are limited;
- off-loading: this is the second challenge, which involves the transfer of tasks between entities in a fog network or from IoT devices to the fog layer. This also needs to ensure the low latency characteristics;
- heterogeneity: maintaining communication between devices which are heterogeneous in nature is another challenge in fog computing; and
- security is another challenge to be considered as more attacks are possible in the fog environment.

8.8 CONCLUSION

The IoT is a group of devices working together and for monitoring, sensing, actuating, and communication purposes. The impact of the IoT in healthcare is real-time remote monitoring: the patient is provided with multiple devices which facilitate real-time monitoring of them. The connected devices also transmit the data from patient to the doctor, hence it reduces the time taken by the patient visiting the hospital or doctor. CVD is one of the most serious issues that needs to be handled efficiently for the betterment of humans. In this chapter, we reviewed the various research work carried out pertaining to smart healthcare systems for CVD. Also, we implemented the random forest and KNNML algorithms. From the results, it is inferred that the random forest algorithm yields better accuracy than KNN. We also reviewed the role of edge/fog computing in IoT healthcare systems. Implementation of smart healthcare systems with efficient ML algorithms incorporating edge/fog computing technologies could be considered for future work.

REFERENCES

[1] *"Cardiovascular Disease,"* Available at: https://www.who.int/health-topics/cardiovascular-diseases/#tab=tab_1.

[2] E.J. Benjamin, P. Muntner, A. Alonso et al., "Heart Disease and Stroke Statistics-2019 Update: A report from the American Heart Association," *Circulation*, 139(10), e56–e528, 2019.

[3] *"Cardiovascular Diseases (CVDs) – Global Fact and Figures,"* Available at: https://www.world-heart-federation.org/resources/cardiovascular-diseases-cvds-global-facts-figures/

[4] A. Khanna and S. Karur, "Evolution of Internet of Things (IoT) and its significant impact in the field of Precision Agriculture," *Computer and Electronics in Agriculture*, 157, 218–231, 2019.

[5] A. Rehman K. Mehmood, and A. Baksh, *"Communication Technology That Suits IoT – A Critical Review," 1st Conference WSN4DC*, Jamshoro, Pakistan, 14–25, 2013.

[6] *"Best Use of Wireless eIoT Communication Technology,"* Available at: https://industry-today.com/best-uses-of-wireless-iot-communication-technology/.

[7] *"NFC Applications: Usecases for NFC Technology,"* Available at: https://info.hidglobal.com/2015-01-NAM-WP-PPC-NFC-Tags-and-Solutions_NFC-Tags-Applications_Request.html?ls=PPC&utm_detail=2014-04-Global-WP-NFC-Tags-and-Solutions-IDT-APAC-EN-321&gclid=CjwKCAjwpqv0BRABEiwA-TySwdSuiCJ2SV8t3co9muVBPuvVtCP07Ygt-w9J18l3NlGWFF6BwOD-ZFxoC4jUQAvD_BwE

[8] *"Applications of Z-Wave Technology,"* Available at: https://www.rfpage.com/applications-of-z-wave-technology/.

[9] S. M. Hassan et al., "Applications of Wireless Technology for Control: A Wireless HART Perpective," *Procedia Computer Science*, 105, 240–247, 2017.

[10] *"RFID Applications,"* Available at: https://www.fibre2fashion.com/industry-article/3271/rfid-applications.

[11] *"What Is 6LOWPAN – The Basics,"* Available at: https://www.electronics-notes.com/articles/connectivity/ieee-802-15-4-wireless/6lowpan.php.

[12] *"RF Communication – Protocol & Application,"* Available at: https://www.elprocus.com/rf-communication-protocol-application/.

[13] *"Top 4 Applications of Zigbee Wireless Technology,"* Available at: https://tvsnext.io/blog/top-4-applications-of-zigbee-wireless-technology.

[14] *"Lowpower Widearea Technology,"* Available at: https://www.gemalto.com/iot/resources/innovation-technology/low-power-wide-area-technology

[15] *"Applications of WIFI,"* Available at: http://ecomputernotes.com/computernetworkingnotes/communication-networks/wifi-applications.

[16] *"Bluetooth Technology and Applications,"* Available at: https://www.itu.int/en/ITU-D/Regional-Presence/AsiaPacific/SiteAssets/Pages/Events/2017/Oct2017CIIOT/CIIOT/8.Session3-3%20Bluetooth%20Technology%20and%20Applications-%E6%9D%A8%E6%B3%A2V3.pdf

[17] S. Arab, H. Ashrafzadeh, and A. Alidadi, "Internet of Things: Communication Technologies, Features and Challenges," *International Journal of Engineering Development and Research*, 6(2), 733–742, 2018.

[18] A. Roger, G. Reza, and G. Hamid, "Smart Grid the Future of the Electric Energy System," *IEEE Transactions*, 1–11, 2018.

[19] E.I. Davies and V. I. E. Anrich, "Design and Implementation of Smart Home System Using Internet of Things," *Journal of Digital Innovations & Contemp Res. in Sci, Eng & Tech*, 7(1), 33–42, 2019.

[20] T. Vairam and S. Sarathambekai, "Proficient Smart Trash can Management Using Internet of Things and SDN Architecture Approach," *International Journal of Enterprise Network Management*, 10 3/4, 241–252, 2019.

[21] N. Loganathan, K. Lakshmi, and P. S. Mayurappriyan, "Smart Energy Management Systems: a Literature Review," *MATE Web of Conferences*, 225(1), 504–508, 2018.

[22] T. Vairam and R. Prasanth, "IoT Based Smart Irrigation and Control System," *International Journal of Creative Research Thoughts*, 6(2), 1677–1681, 2018.

[23] *"Classifications - Machine Learning Algorithm,"* Available at: https://www.simplilearn.com/classification-machine-learning-tutorial.

[24] S. Mohan Kumar and D. Majumder, "Helathcare Solution Based on Machine Learning Applications in IoT and Edge Computing," *International Journal of Pure and Applied Mathematics*, 119(16), 1473–1484, 2018.

[25] Fonseca et al., "Deep Learning and IoT to Assist Multimorbidity Home Based Healthcare," *Journal of Health & Medical Informatics*, 8(3), 1–4, 2017.

[26] *"Top 10 Applications of Machine Learning in Healthcare,"* Available at: https://www.flatworldsolutions.com/healthcare/articles/top-10-applications-of-machine-learning-in-healthcare.php.

[27] I. Iswanto, L. Laxmi, S. Nguyen, P. Thanh, H. Wahidah, and M. Andino, "Identifying Diseases and Diagnosis Using Machine Learning," 8, 978–981, 2019. 8.35940/ijeat.F1297.0886S219.

[28] G. S. Fu et al., "Machine Learning for Medical Imaging," *Journal of Healthcare Engineering*, 2019, 1–2, 2019.

[29] J. Vamathevan et al., "Applications of Machine Learning in Drug Discovery and Development," *National Review Drug Discovery*, 18(6), 463–477, 2019. doi: 8.1038/s41573-019-0024-5.

[30] M. Uddin, Y. Wang, and M. Woodbury-Smith, "Artificial Intelligence for Precision Medicine in Neurodevelopmental Disorders," *NPJ Digit Medicine*, 2, 112 2019, https://doi.org/8.1038/s41746-019-0191-0.

[31] M. Ani et al., "Machine Learning Based Behavioral Modification," *International Journal of Engineering and Technology*, 8, 1134–1138, 2019, 8.35940/ijeat.F1299.0886S219.

[32] K. Abhishek, S. Tvm, and D. Vishal, Machine Learning Implementation for Smart Health Records: A Digital Carry Card, *Global Journal on Innovation, Oppurtunities and Challenges in AAI and Machine Learning*, 3 1 12–22, 2019.

[33] Z. Rayan, M. Alfonse, A. Badeeh, and M. Salem, "Machine Learning Approaches in Smart Health," *Procedia Computer Science*, 154, 361–368, 2019.

[34] P. Shah, F. Kendall, S. Khozin et al., "Artificial Intelligence and Machine Learning in Clinical Development: A Translational Perspective," *NPJ Digital Medicine*, 2, 69, 2019, https://doi.org/8.1038/s41746-019-0148-3.

[35] K. Wazny, "Applications of Crowdsourcing in Health: An Overview," *Journal of Glob Health*, 8(1), 010502, 2018. doi: 8.7189/jogh.08.010502.

[36] M. Müller Martin and S. Marcel, "Crowdbreaks: Tracking Health Trends Using Public Social Media Data and Crowdsourcing," *Frontiers in Public Health*, 7, 81–87, 2019, 8.3389/fpubh.2019.00081.

[37] D. Jarrett, E. Stride, K. Vallis, and M. J. Gooding, "Applications and Limitations of Machine Learning in Radiation Oncology," *An International Journal of Radiology, Radiation Oncology and All related Science*, 92(1100), 1–12, 2019.

[38] M. Feng, G. Valdes, N. Dixit, and T. D. Solberg, "Machine Learning in Radiation Oncology: Opportunities, Requirements, and Needs," *Frontiers in Oncology*, 8, 18, 2018, Published 2018 Apr 17, doi: 8.3389/fonc.2018.00110.

[39] V. Sharma, "Malaria Outbreak Prediction Model Using Machine Learning," *International Journal of Advanced Research in Computer Engineering & Technology*, 4(12), 4415–4419, 2015.

[40] P. Verma and S. K. Sood, "Fog Assisted-IoT Enabled Patient Health Monitoring in Smart Homes," *IEEE Internet of Things Journal*, 5(3), 1789–1796, 2018.

[41] S. K. Sood and I. Mahajan, "IoT-Fog-Based Healthcare Framework to Identify and Control Hypertension Attack," *IEEE Internet of Things Journal*, 6(2), 1920–1927, 2019.

[42] S. Rani, S. H. Ahmed, and S. C. Shah, "Smart Health: A Novel Paradigm to Control the Chikungunya Virus," *IEEE Internet of Things Journal*, 6(2), 1306–1311, April 2019.

[43] R. Ding, H. Zhong, J. Ma, X. Liu, and J. Ning, "Lightweight Privacy-Preserving Identity-Based Verifiable IoT-Based Health Storage System," *IEEE Internet of Things Journal*, 6(5), 8393–8405, 2019.

[44] G. Yang et al., "IoT-Based Remote Pain Monitoring System: From Device to Cloud Platform," *IEEE Journal of Biomedical and Health Informatics*, 22(6), 1711–1719, Nov. 2018.

[45] A. F. Hussein, N. A. Kumar, M. Burbano-Fernandez, G. Ramírez-González, E. Abdulhay, and V. H. C. DeAlbuquerque, "An Automated Remote Cloud-Based Heart Rate Variability Monitoring System," *IEEE Access*, 6, 77055–77064, 2018.

[46] S. Vadrevu and M. S. Manikandan, "Real-Time Quality-Aware PPG Waveform Delineation and Parameter Extraction for Effective Unsupervised and IoT Health Monitoring Systems," *IEEE Sensors Journal*, 19(17), 7613–7623, Sept. 1, 2019.

[47] J. B. Bathilde, Y. L. Then, R. Chameera, F. S. Tay, and D. N. A. Zaidel, "*Continuous Heart Rate Monitoring System as an IoT Edge Device*," *2018 IEEE Sensors Applications Symposium (SAS)*, Seoul, 1–6, 2018.

[48] K. C. Tan, E. J. Teoh, Q. Yu, and K. C. Goh, "A Hybrid Evolutionary Algorithm for Attribute Selection in Data Mining," *Expert Systems with Applications*, 36(4), 8616–8863, 2014.

[49] A. F. Otoom, E. E. Abdallah, Y. Kilani, A. Kefaye, and M. Ashour, "Effective Diagnosis and Monitoring of Heart Disease," *International Journal of Software Engineering and Its Applications*, 9(1), 143–s156, 2015.

[50] G. Parthiban and S. K. Srivatsa "Applying Machine Learning Methods in Diagnosing Heart Disease for Diabetic Patients," *International Journal of Applied Information Systems (IJAIS)*, 3(7), 25–30, 2012.

[51] V. Chaurasia and S. Pal, "Data Mining Approach to Detect Heart Diseases," *International Journal of Advanced Computer Science and Information Technology (IJACSIT)*, 2, 56–66,2014

[52] A. Malav, K. Kadam, and P. Kamat, "Prediction of Heart Disease Using k-Means and Artificial Neural Network as Hybrid Approach to Improve Accuracy," *International Journal of Engineering and Technology*, 9(4), 3081–3085,2017.

[53] R. Ani, S. Krishna, N. Anju, M. S. Aslam, and O. S. Deepa, "*Iot Based Patient Monitoring and Diagnostic Prediction Tool Using Ensemble Classifier*," *2017 International Conference on Advances in Computing, Communications and Informatics (ICACCI)*, Udupi, 1588–1593, 2017.

[54] S. Dash et al., "Edge and Fog Computing in Healthcare - A Review," *Scalable Computing: Practice and Experience*, 20(2), 191–205, 2019.

[55] "*Will Edge Computing Transform Healthcare*," Available at: https://healthtechmagazine.net/article/2019/08/will-edge-computing-transform-healthcare.

[56] S. Oueida et al., "An Edge Computing Based Smart Healthcare Framework for Resource Management," *Sensors*, 18(12), 1–22, 2018.

[57] C. Mahulea et al., "Modular Petri net Modeling of Healthcare Sytems," *Flexible Services and Manufacturing Journal*, 30, 329–357, 2018.

[58] M. Chen et al., "Edge Cognitive Computing Based Smart Healthcare System," *Future Generation Computer Systems*, 86, 403–411, 2018.

[59] *"IoT in Healcare : Benefits, Challenges and Applications,"* Available at: https://www.valuecoders.com/blog/technology-and-apps/iot-in-healthcare-benefits-challenges-and-applications/.

[60] M. Mukherjee, L. Shu, and D. Wang, "Survey of Fog Computing: Fundamental, Network Applications, and Research Challenges," *IEEE Communications Surveys Tutorials*, 1–1, 20(3), 1826–1857, 2018.

[61] S. Sarkar, S. Chatterjee, and S. Misra, "Assessment of the Suitability of Fog Computing in the Context of Internet of Things," *IEEE Transactions on Cloud Computing*, 6(1), 46–59, 2018.

[62] K. Intharawijitr, K. Iida, and H. Koga, *"Analysis of Fog Model Considering Computing and Communication Latency in 5G Cellular Networks,"* *2016 IEEE International Conference on Pervasive Computing and Communication Workshops, PerCom Workshops 2016*, Sydney, Australia, 2016.

[63] N. Abbas, Y. Zhang, A. Taherkordi, and T. Skeie, "Mobile Edge Computing: A Survey," *IEEE Internet of Things Journal*, 5(1), 450–465, 2017.

[64] H. Dubey and N. P. Constant, "Enhancing Telehealth Big Data through Fog Computing," *CoRR*, abs/1605.09437, 2015.

[65] M. Al-Khafajiy et al., *"Towards Fog Driven IoT Healthcare: Challenges and Framework of Fog Computing in Healthcare,"* *ICFNDS'18*, 2018.

Index

Page numbers in *italics* refer to content in *figures*; page numbers in **bold** refer to content in **tables**.

For Product Safety Concerns and Information please contact our EU
representative GPSR@taylorandfrancis.com
Taylor & Francis Verlag GmbH, Kaufingerstraße 24, 80331 München, Germany

9 780367 636944